General Preface to the Series

It is no longer possible for one textbook to cover the whole field of Biology and to remain sufficiently up to date. At the same time teachers and students at school, college or university need to keep abreast of recent trends and know where the most significant developments are taking place.

To meet the need for this progressive approach the Institute of Biology has for some years sponsored this series of booklets dealing with subjects specially selected by a panel of editors. The enthusiastic acceptance of the series by teachers and students at school, college and university shows the usefulness of the books in providing a clear and up-to-date coverage of topics, particularly in areas of research and changing views.

Among features of the series are the attention given to methods, the inclusion of a selected list of books for further reading and, wherever possible, suggestions for practical work.

Reader's comments will be welcomed by the author or the Education Officer of the Institute.

1976

The Institute of Biology
41 Queens Gate,
London, SW7 5HU

Preface

Because of their widespread importance, and because of the ease with which they can be maintained in the laboratory, the biology of locusts has been more extensively studied than that of any other insect. In this book some attempt is made to bring together the field and laboratory studies and to put them in the context of control with its practical and organizational problems. My aim has been to indicate the major areas of interest and study; inevitably in covering such a wide field the treatment is superficial, but hopefully the book will serve to make students aware of, and encourage them to take an interest in, this unique story. To help in maintaining a coherent account, but also reflecting my own background, emphasis is placed on the three locust species most fully studied by scientists in Africa and the Middle East. Ecologically some of the other locust species are equally well studied and are equally as fascinating in their biology. A working knowledge of insect anatomy and general terminology has been assumed. I am indebted to many colleagues, but especially to Dr E. Bernays, Dr P. Ellis, Dr P. Haskell, Mr C. Hemming, Dr P. Hunter-Jones and Mr J. Roffey for their constructive comments.

Centre for Overseas Pest Research,   R. F. C.
London, 1976

Contents

1 What are Locusts?

1.1 Introduction

In the ruins of Nineveh, excavated during the last century, the 'noxious locust' is recorded on tablets compiled during the ninth century BC from information accumulated much earlier, possibly 1000 years before, in Sumeria. At about the same time the Shang Kingdom in China appointed antilocust officers and by AD 720 a system for forecasting locust plagues was established in China. Locusts were a problem to earlier civilizations just as they are today, but what are locusts and why are they such pests?

Locusts look just like large grasshoppers and are not morphologically distinguishable from them. They differ in their behaviour. When locusts are present in large numbers they become gregarious—that is, they tend to be attracted to each other and to group together rather than remaining dispersed through the environment. This can be very simply shown in experiments in which locusts, reared in a crowd, are introduced into an arena which is effectively featureless and in which heating and lighting conditions are uniform (Fig. 1–1). Within 30 minutes the insects have grouped together and if they are disturbed so as to separate them they soon come together again. If the same experiment is repeated with a grasshopper, such as *Cyrtacanthacris tatarica*, the insects remain scattered round the arena and no more grouping occurs than would be expected in a random distribution. But the distinction is not clear-cut, and if we look at a range of species, we find that there is a series, from species that do not group at all to the desert locust which groups very strongly (Fig. 1–1).

The second major distinguishing feature of locusts is that they migrate during the daytime in swarms containing many millions of individuals. Migration itself is not a peculiarity of locusts because there is an increasing amount of evidence that grasshoppers also migrate, but only as isolated individuals and at night. Extensive day flights by grasshoppers have not been recorded, except for one or two species which occasionally form swarms, such as *Melanoplus sanguinipes* in North America and *Aiolopus savignyi* in the Sudan and West Africa.

Locusts do not always occur in these vast, migrating aggregations; they also exist as solitary individuals. In this solitary phase (p. 40) they show no tendency to aggregate when put together in a crowd (Fig. 1–1) and with these solitary individuals the distinction between locusts and grasshoppers breaks down completely. We are forced to conclude that even on behavioural grounds there is no clear cut distinction between locusts and grasshoppers. Rather we must think of them as forming a series, from grasshoppers which never aggregate on the one hand, to the locusts which very commonly do so on the other. In between is a range of species which sometimes aggregate, but sometimes do not. If they do so

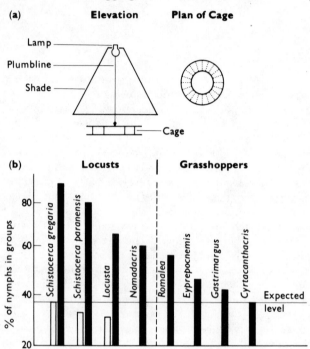

Fig. 1–1   (a) Apparatus used to measure the grouping behaviour of locusts and grasshoppers. (b) Grouping behaviour of locusts and grasshoppers tested in the apparatus shown in (a). For each test, 30 insects of one species were introduced into the apparatus and the percentage in groups of two or more was recorded after 30 minutes. Black bars = insects reared in crowds; white bars = insects reared in isolation. The horizontal line indicates the expected level of grouping if the insects were randomly distributed. (After ELLIS, P., 1962, *Colloq. int. Cent. nat. Rech. sci.* no. 114, 123–43.)

fairly commonly we regard them as locusts, if they do so only occasionally, or not at all, we think of them as grasshoppers.

Neither the tendency to aggregate nor the tendency to migrate would be important if the locusts did not also occur in vast numbers, at least for some of the time. Young locusts may occur in bands, commonly at densities of $100/m^2$ and sometimes exceeding $1000/m^2$, while the bands often exceed $10\,000\ m^2$ in area. Adult swarms may extend over tens of square kilometres and a single swarm may contain more than

1 000 000 000 locusts weighing 1 500 000 kg. Since these insects eat approximately their own weight of vegetation daily (p. 22) it is obvious that they may do an immense amount of damage to pasture or crops. The Food and Agriculture Organization of the United Nations (FAO) has estimated that in 1955, when the last desert locust plague was at its height, crops valued at over £5 000 000 were destroyed. In Morocco alone in late 1954 and early 1955, damage, mainly to citrus, was estimated at £4 500 000. In a technically advanced society losses of this magnitude may be serious; in communities depending on subsistence or peasant farming their effects can be disastrous. This is why locusts are important and why so much time and effort has gone, and is going, into studies of their biology and control.

1.2   Taxonomic position

Locusts are large to medium-sized insects belonging to the order Orthoptera. This order, which includes the crickets and grasshoppers, is characterized by the possession of powerful chewing jaws and, generally, by having the hind legs enlarged for jumping. It is commonly divided into

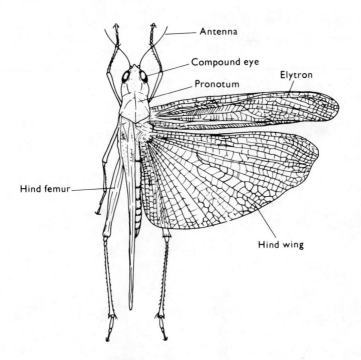

Fig. 1–2   The main external features of an adult locust.

five superfamilies (Table 1), but the locusts belong to only one of these, the Acridoidea, despite the common name of grouse locusts which is sometimes given to the Tetrigoidea.

The Acridoidea is the largest of the superfamilies and includes about 10 000 species, which are all basically very similar in form (Fig. 1–2), and of which about 10 or 12 can be called locusts (Table 2), while the rest are known simply as 'grasshoppers'. The locusts are not differentiated from grasshoppers by any particular morphological characteristics and they do not form a distinct taxonomic group. In fact, they occur in several

Table 1   The superfamilies of Orthoptera

Superfamily	Common names	Characteristics
Tettigonioidea	bush crickets, long-horned grasshoppers	antennae as long as body, with more than 30 segments tympanal organs on fore tibiae tarsi 4-segmented fore-wings generally held in roof-like manner over back female with an elongate, laterally flattened ovipositor
Grylloidea	crickets	antennae as long as body, with more than 30 segments tympanal organs on fore tibiae tarsi 3-segmented forewings generally held flat over back female with an elongate, roughly cylindrical ovipositor
Acridoidea	grasshoppers and locusts	antennae short, less than 30 segments tympanal organs in abdomen tarsi 3-segmented fore-wings held in a roof-like manner over back no elongate ovipositor
Tetrigoidea	grouse locusts	antennae short, less than 30 segments fore- and mid-tarsi 2-segmented, hind-tarsi 3-segmented characterized by elongate pronotum which covers the abdomen and the wings no elongate ovipositor
Tridactyloidea	pygmy mole-crickets	antennae short, with 12 or less segments tarsi 1- or 2-segmented no elongate ovipositor

different subfamilies of the Acridiodea as Table 2 shows, and a locust species commonly shows closer morphological affinities with certain grasshoppers than it does with other locusts. For instance, although there are several locust species in the subfamily Cyrtacanthacridinae, this subfamily also contains a number of typical grasshoppers. The same situation holds in the Oedipodinae where the migratory locust, *Locusta migratoria*, has much greater affinities with the grasshopper *Gastrimargus* than it does with, say, the desert locust, *Schistocerca gregaria*, which is in a different subfamily.

1.3   Distribution

Locusts occur in all the warmer regions of the world (Table 2). The desert locust, *Schistocerca gregaria*, occurs from West Africa to India, and

Table 2   The different species of locusts

Scientific name	Common name	Sub-family	Wing span of female (mm)	Distribution
Schistocerca gregaria	Desert locust	Cyrtacanthacridinae	125	Northern Africa, Arabia, Indian sub-continent, from about 10 to 35°N
Schistocerca americana americana		Cyrtacanthacridinae	116	Central America
Shistocerca americana paranensis		Cyrtacanthacridinae	115	South America
Anacridium melanorhodon	Sahelian tree locust	Cyrtacanthacridinae	143	In a belt across Africa south of the Sahara, from about 10 to 20°N, but extending to the equator in the east
Anacridium wernerellum	Sudanese tree locust	Cyrtacanthacridinae	128	In a zone slightly further south than *A. melanorhodon*
Nomadacris septemfasciata	Red locust	Cyrtacanthacridinae	131	In Africa, mainly south of the equator to about 30°S
Patanga succincta	Bombay locust	Cyrtacanthacridinae	138	South-west Asia
Melanoplus spretus	Rocky mountain locust	Catantopinae	43	North America, now said to be extinct
Chortoicetes terminifera	Australian plague locust	Oedipodinae	58	Australia
Locusta migratoria	Migratory locust	Oedipodinae	101	Different subspecies in Southern Europe, Africa south of the Sahara, Malagasy Republic, Southern Russia, China, Japan, Philippines, Australia
Locustana pardalina	Brown locust	Oedipodinae	93	South Africa
Dociostaurus maroccanus	Moroccan locust	Gomphocerinae	67	Middle East and Mediterranean countries

extends north to Iran and south to Kenya. The migratory locust, *Locusta migratoria*, is even more widespread extending through Africa eastwards to Japan, the Philippines and Australia. It has been separated into a number of different subspecies depending on which region it comes from, but at least some of the subspecies interbreed in the laboratory. No species is common to both America and Africa and some, like the brown locust in South Africa, are relatively restricted in their distribution.

In this account of locust biology more attention is paid to the three main African locust species, the desert locust (*Schistocerca gregaria*) the migratory locust (*Locusta migratoria*) and the red locust (*Nomadacris septemfasciata*) than to other species. This partly reflects the greater amount of work on these species and partly their more extensive migrations which give the insects considerable international importance.

For a comprehensive account of Locust biology see UVAROV, 1966. For information on the different species and their control see *The Locust Handbook*.

2 The Life History of a Locust

Although locusts are not confined to a single taxonomic group the development of all the species is basically similar and in the following account *Locusta migratoria* is taken as an example. Reference to other species is made where they differ from *Locusta*.

2.1 The egg

The egg of *Locusta* is a slightly curved, sausage-shaped structure about 6 mm long. The outer shell of the egg, the chorion, has a hexagonal pattern which indicates the outlines of the follicle cells which secreted it. The precise form of this patterning varies from species to species. Close to one end of the egg are about 40 openings which lead through the chorion. These are the micropyles and it is through them that spermatozoa reach the surface of the oocyte to effect fertilization. The eggs are laid in a group or clutch up to 15 cm below the surface of the ground. They are regularly arranged (Fig. 2–1), although the arrangement varies in different species, and are held together by a rather crisp, frothy material. This also coats the

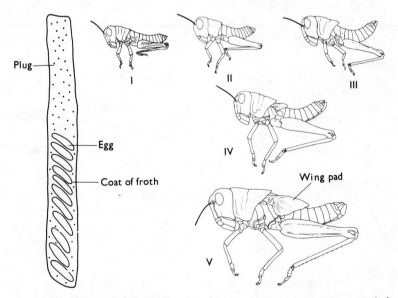

Fig. 2–1 An egg-pod showing the arrangement of eggs, and the five nymphal instars of *Locusta migratoria*.

outside of the egg-clutch and continues as a solid plug from the top of the egg-clutch to the soil surface. The whole structure is known as an egg-pod.

The froth is a product of the glands which form the walls of the oviducts of the female. It consists of tanned protein. Where it surrounds the eggs it reduces water loss to some extent and the plug forms an easy route by which the newly hatched locusts can reach the surface. But whether or not these are the main functions of the froth is not known.

The number of eggs in a pod is roughly equivalent to the number of ovarioles in the female ovaries because only one oocyte at a time matures in each ovariole. In practice, it is uncommon for all the ovarioles to produce oocytes so that the number of eggs laid is nearly always less than the potential maximum. Sometimes many of the ovarioles fail to function and relatively few eggs are laid (see p. 51). *Locusta* has about 100 ovarioles and in the laboratory commonly lays pods containing 70 eggs, but with poor food the number may be reduced to 30. The number of eggs in a pod varies with the species of locust (see Table 3, p. 43), generally in proportion to the size of the insect, and also with the degree of crowding which the insects have experienced.

2.2 Oviposition

Eggs are usually laid in areas of bare ground close to vegetation. Females will not oviposit in dry ground or in waterlogged soil, but apart from these extremes will lay in soils with widely differing moisture contents. Commonly a final decision on whether or not to lay is made only after the female has made a hole with her abdomen so that if there is moisture below the surface a female may oviposit even though the surface of the ground is quite dry. Saline soils are avoided. In the laboratory, females also exhibit preferences for soils with certain particle sizes but how important this behaviour is in the field is impossible to say. Probably the choice of oviposition site depends largely on where the female happens to be when she is ready to lay. Only if the soil is positively unsuitable will she move to another site.

A feature of locusts in swarms is that they tend to oviposit close together (Fig. 2–2). This arises largely from the reactions of the locusts to each other, and in the desert and migratory locusts a pheromone is produced which tends to keep the females in a group. There is little evidence that they are attracted from a distance and the pheromone is detected by contact rather than by smell. It is produced by nymphs as well as by adults.

The mechanism of oviposition is of particular interest because the female in digging the hole for her eggs approximately trebles the length of her abdomen. In the normal female the abdomen is about 3 cm long, but in the course of oviposition it can be extended to as much as 13.5 cm.

When the female starts to dig she arches her abdomen so that the valves of the ovipositor are vertical and touching the ground. By opening and closing the valves she digs into the soil, forcing the particles outward and upwards. Periodically she stops and revolves the abdomen about its axis,

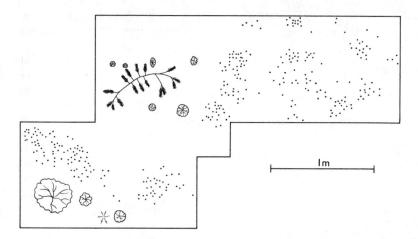

Fig. 2–2   Distribution of egg-pods of *Schistocerca gregaria* in part of a laying site showing the clumping of the egg-pods even in areas of bare ground. Each dot represents one egg-pod and each plant in the area is shown, different symbols indicating different species. (After STOWER, W. J., POPOV, G. B. and GREATHEAD, D. J., 1958, *Anti-Locust Bull.* no. 30.)

before starting to dig again. As she digs deeper the intersegmental membranes become unfolded, because the hard sclerites of the abdominal segments are inextensible, but the membranes between segments 4–5 and 5–6 extend much more than the others, until in the fully extended abdomen they may each be over 1 cm long (Fig. 2–3).

Stretching also involves the internal organs such as the alimentary canal, nervous system and intersegmental muscles. In the normal insect the first two are slightly convoluted, but nevertheless they must be considerably stretched. Even more remarkable is the extension of the intersegmental longitudinal muscles.

The extension of the abdomen is achieved mechanically by the pull of the ovipositor valves, but the extension is maintained by pressure from within the body. As the abdomen elongates the volume of the body increases, but the pressure is maintained by expansion of air sacs and by swallowing air. Air is pumped into the air sacs by vigorous movements of the head and as the eggs are laid more air sacs are expanded, so that at the end of oviposition they fill the cavities of the first five abdominal segments. Having laid her full complement of eggs, the female slowly

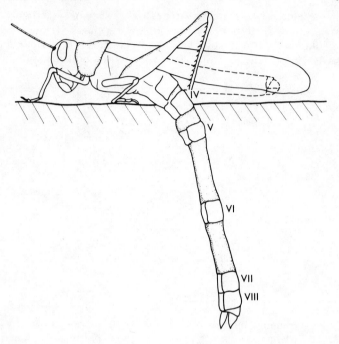

Fig. 2–3   Diagram showing the extension of the abdomen of *Locusta* during oviposition. Membranous areas are hatched. Notice the enormous extension between segments V and VI, and VI and VII. The normal size and position of the abdomen is indicated by the broken line.

withdraws her abdomen, filling the hole with froth as she does so. The whole process of oviposition is complete in about two hours.

2.3   Embryonic development

When the egg is first laid it is one big cell and the first stages of embryonic development involve nuclear division without cell division. Subsequently the nuclei migrate to the periphery of the egg and a layer of cells, known as the blastoderm, is formed surrounding the yolk. Some larger cells, vitellophages, remain within the yolk apparently aiding in its digestion. The embryo proper first appears as a disc of cells at the micropylar end of the egg. The disc soon becomes a hollow ball of cells and the embryo develops on the inner surface. It elongates towards the opposite end of the egg and the future head end broadens. At the same time the blastoderm secretes a layer of cuticle, known as the serosal cuticle, towards the outside. Later the embryo becomes segmented and the appendages appear. The mandibles, maxillae and labial palps are first seen as appendages serially homologous with the legs, a convincing demonstration of their derivation from leg-like appendages. Subsequently they become fused with the unsegmented anterior part to form the head.

About halfway through the period of embryonic development the embryo moves round the end of the egg so that it comes to lie with its head pointing in the opposite direction. The result of this movement, known as katatrepsis, is to bring the yolk effectively within the body of the embryo; previously the embryo floated on the surface of the yolk. Subsequently the embryo elongates to occupy the full length of the egg, coming to look more and more like a small locust. On day 7, if the eggs are incubated at 33°C, another layer of cuticle, known as the embryonic cuticle, is produced, but this soon separates from the underlying epidermis and the first instar cuticle is laid down underneath it. The embryonic cuticle is not shed until the locust hatches.

In *Locusta* and all the other locust species the eggs take up water rapidly from the surrounding soil before katatrepsis, approximately doubling their weight in doing so (Fig. 2–4). Development does not continue if sufficient water is not available for this and in very dry soil the eggs slowly lose water. If moisture subsequently becomes available within a few days the water content of the eggs is restored and the only effect of the shortage is to delay development. More prolonged lack of water leads to death,

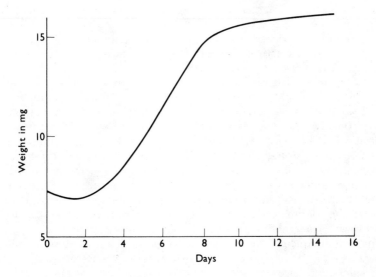

Fig. 2–4 Changes in weight of an egg of *Locusta* between oviposition (day 0) and hatching (day 16) at 27°C. The increase in weight is due to the uptake of water. (After SHULOV, A. and PENER, M. P., 1959, *Locusta*, **6**, 73–88.)

although the diapausing eggs of *Locustana* and *Chortoicetes* (see below) are better fitted to survival than the rest.

In most locust species development is continuous and in *Locusta* the first instar nymphs hatch about 11 days after the eggs are laid if they are incubated at 33°C, but in *Locustana* and *Chortoicetes* there is a delay in embryonic development associated with survival over a dry period. In *Locustana* development stops just before the uptake of water and katatrepsis occur. This state of diapause may persist for over three years and enables the species to survive long periods of drought which occur in its habitat. During this time the rate of water loss is very low compared with that in a non-diapausing species.

2.4 Hatching

Although the free-living locust has no appendages on the abdomen apart from the genitalia and cerci, the rudiments of appendages can be seen on all the abdominal segments at an early stage of embryonic development. Subsequently these are resorbed except for those on the terminal segments, which contribute to the genitalia, and for those on the first segment which persist as special structures known as pleuropodia. A little over a day before hatching, the cells of the pleuropodia secrete an enzyme into the extra-embryonic fluid. This enzyme digests the inner part of the serosal cuticle which forms the main barrier to hatching since the chorion tends to fragment as the egg swells when its water content increases. When it is ready to hatch the larva swallows the extra-embryonic fluid and by a series of complex body movements, splits the weakened serosal cuticle, usually transversely over the neck region. The nymph is then able to wriggle its way out of the egg.

At the time of hatching the locust is still enclosed within the embryonic cuticle and inside this the first instar cuticle is well formed, although it is still soft and flexible apart from some specialized structures such as spines and joints. At this stage the locust is known as a vermiform larva because its movements are worm-like, and it wriggles its way to the surface of the soil, usually through the frothy plug. Movement through the froth is aided by a pair of eversible swellings called ampullae on the membrane of the neck which are forced out by haemolymph pressure. With them the insect anchors its head in the froth or soil through which it is moving. Then the abdomen is drawn up and the insect obtains a new purchase with the tip of its abdomen on some point on the side of the burrow it has made. Subsequently the ampullae are withdrawn and the head is forced forwards into the substratum. By a series of such movements the insect reaches the surface (Fig. 2–5).

As soon as the larva reaches the surface it starts to swallow air. The embryonic cuticle is drawn backwards so that it is very taut over the head and finally it splits. Then it is worked back over the body and is kicked off by the hind legs. The cast embryonic cuticles are visible as small white particles, about 1 mm across, on the surface of the ground round the exit hole through which the insects emerged.

Having reached the surface the insect rapidly hardens and darkens so

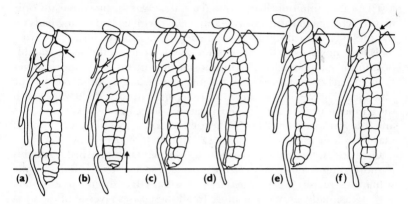

Fig. 2–5   The changes in body form of a vermiform larva of *Schistocerca* as it wriggles its way to the surface after escaping from the egg. The positions of two sand grains are indicated next to the head. (**a**) The larva obtains a purchase by inflating the ampullae (arrow). (**b**) The abdomen is drawn up (arrow). (**c**) The tip of the abdomen is flexed to obtain a purchase and the anterior part of the larva extends (arrow) as the ampullae are drawn in. (**d**) and (**e**) Elongation continues and the head is pushed up past the sand grains. (**f**) The ampullae are inflated and the sequence begins again. (After BERNAYS, 1971.)

that within about two hours it has the appearance of a typical but small locust. Darkening is controlled by a blood-borne factor which is released just as the insect splits the embryonic cuticle.

For a detailed account of the processes involved in hatching see BERNAYS, 1971.

2.5   Nymphal development

Larval or nymphal development follows the usual hemimetabolous pattern. The term larva emphasises the underlying similarity of the physiological processes in hemi- and holo-metabolous insects; the early developmental stages of locusts may with equal justification be termed nymphs in view of their superficial morphological resemblance to the adults. They are called nymphs in this book because this is the term commonly used in literature. The general development of all species is similar and again *Locusta* may be taken as an example.

The first instar stage lasts for about five days at 33°C and then the insect moults to the next stage. At each ecdysis the insect becomes progressively larger and, in terms of size of head capsule or length of leg, growth takes place in a series of steps. In terms of weight, however, the increase in size is more or less progressive, though there is a slight loss of weight at each ecdysis due to the shedding of the old cuticle.

There are commonly five nymphal instars. Over the first three, the only marked change in form involves some enlargement of the wing pads on the meso- and meta-thoracic segments. In the fourth instar, the wing pads twist upwards so that they point over the back of the insect and this position is maintained in the fifth instar in which they are much larger than in earlier instars. The forewing pad is largely obscured by the hindwing (Fig. 2–1). The period of time spent in each instar varies with the temperature, but at 33°C, the durations of the five instars are 5, 4, 4, 5 and 8 days respectively. At 24°C the total period of nymphal development may be as long as 45 days while at 42°C it is reduced to 20 days.

Occasionally *Locusta* may pass through an extra nymphal instar and this is more often true of *Schistocerca gregaria*, especially of females in the solitarious phase (p. 41). *Nomadacris* normally has six nymphal instars with occasionally as many as nine. In all these cases reversal of the wing pads over the back of the insect occurs at the moult into the penultimate nymphal instar. In some species, although not in *Locusta*, a dark stripe appears in the eye at each moult so that it is possible to determine the stage of development from the number of eye stripes. At the final moult the adult emerges, the wings taking about 20 minutes to expand to their full extent.

2.6 Adult maturation

When the locust emerges as an adult insect it is not sexually mature; maturation takes a further week or more, depending on the species and the prevailing conditions. For about the first ten days after the final moult the insect undergoes a period of somatic growth in which the flight muscles increase in size and become fully functional and fat is stored in the fat body. During this period the insect increases steadily in weight and the cuticle thickens and becomes harder.

Subsequently the insect becomes sexually mature. In the female the ovaries are already fully developed at the time of the final moult, but the oocytes are very small. No further development takes place until after the period of somatic growth. At this time specific proteins are found in the haemolymph probably having been synthesized in the fat body. These are taken up by the terminal oocytes in each ovariole (the terminal oocyte is the oocyte nearest the oviduct) which become yellow and rapidly increase in size until within a few days they are fully developed and the chorion is laid down by the follicle cells.

The process of maturation in the female is controlled by the median neurosecretory cells in the brain and by the corpora allata, which become functional again in the adult, having ceased to produce hormone for a period before and after the final moult. Maturation can be speeded up by implanting corpora allata into immature females, which, in the case of *Locusta*, can then be induced to oviposit 10–14 days after moulting compared with the usual 28–33 days. Conversely, removal of the corpora allata prevents the females from maturing at all, although they may live for very long periods.

The precise timing at which maturation occurs under natural conditions varies with the species, and to some extent with environmental conditions. This variability is regulated via the neurosecretory/corpora allata system. In *Locusta* maturation usually follows the period of somatic growth, provided adequate food is available for protein synthesis. To some extent in *Locusta*, but more markedly in *Schistocerca*, maturation is affected by other locusts. Immature females and males become mature much more quickly if they are in the presence of mature males. It has been shown that this is due to a pheromone produced by the males, probably from epidermal cells. Continuous association with very young immature locusts retards maturation, possibly due to an inhibiting pheromone. The effect of these pheromones is to bring about some degree of synchrony in the time of maturation, and hence of oviposition, of large numbers of insects in a swarm.

It is known that in *Schistocerca gregaria*, maturation may be delayed for some months under dry or cool conditions, but may be stimulated by the odours of certain plants. These odours are produced by the newly developing foliage of plants such as myrrh which are stimulated to grow by rain. This has obvious adaptive value to the insect in stimulating maturation at a time when food for the subsequent generation of nymphs is most likely to be present.

Finally, there are species such as *Nomadacris* in which a long delay between adult emergence and maturation is typical. In this case the delay may be regarded as an adult diapause in which the insects survive the adverse conditions of a very severe dry season. During this period, they require only sufficient food for body maintenance, while the ability to fly enables them to make local movements in search of food.

Diapause is often regulated by changes in day length. This is true also of *Nomadacris* but since its range lies across the equator the changes in day length which it experiences are only slight. It has been found that day length is effective during nymphal development as well as in the adult and that diapause is induced if the nymphs and early adults experience day lengths decreasing from 13 to 12 hours. This is the situation in the outbreak areas (see p. 46). The end of diapause is associated with the beginning of the rainy season, but maturation may begin before any rain falls and the exact nature of the relationship is not known.

Maturation in male locusts also involves the corpora allata and the median neurosecretory cells, but their importance varies in different species.

For pheromones affecting maturation see NORRIS, 1968. For plant chemicals affecting maturation see CARLISLE, ELLIS and BETTS, 1965. For control of maturation in male locusts see PENER, GIRARDIE and JOLY, 1972.

 2.7 Mating

Mating in locusts does not involve any special courtship. In most species a mature male simply jumps onto the back of a female. If she is receptive she tolerates his presence and the genitalia are coupled, otherwise the male is kicked off. The insects may remain coupled for long periods and during this time sperm are transferred from the male to the female.

Sperm transfer involves the production of a spermatophore by the male. Production of the spermatophore only begins after the genitalia of the pair are coupled. In *Locusta* a series of different secretions is produced by the male accessory glands and a tube is formed which is forced out into the spermathecal duct of the female, reaching almost to the spermathecal sac. Through this tube sperm are forced into the spermatheca. In this case the bulk of the spermatophore remains within the male genitalia so that when the pair separate the tubular part entering the female duct is broken off. This is partially digested and eventually expelled from the duct some days later. For a part of this time further matings are unsuccessful, apparently because the remains of the tube block the duct. The remains of the spermatophore in the male are ejected soon after separation.

In *Schistocerca* the spermatophore is much simpler, consisting of a bulb with only a short tube entering the spermathecal duct. The whole structure is transferred to the female and in the course of a single copulation, lasting perhaps four hours, up to 14 spermatophores may be produced. When the pair separate these remain attached to the female, but are soon discarded.

It is important to separate clearly between the process of *insemination* of the female, that is the transfer of sperm from the male to the spermatheca, and the process of *fertilization* of the egg which may occur days or weeks later. In the interim period sperm are stored in the female spermatheca. It is known that a single insemination will serve to fertilize all the eggs which a female lays during her life. This may amount to as many as 10 clutches though in the field the number is probably only two or three. Nevertheless multiple mating commonly occurs. In this case it appears that the last sperm to be received by the female are most effective in fertilizing the eggs. This has been shown by making successive matings with normal and albino males. It is also known that mating stimulates oogenesis via the neurosecretory cells so that the interval between successive ovulations is reduced.

3 Feeding Behaviour

Locusts are economically important by virtue of the damage which they do by feeding, so it is necessary to have an understanding of their feeding behaviour.

3.1 The sensory system involved in feeding

Locusts locate and identify their food using the visual and chemical senses. The compound eyes are large (Fig. 1–2) and made up of many ommatidia just as in other insects. In adult *Schistocerca* there are about 9400 ommatidia in each eye.

The chemoreceptors concerned with feeding are located on the antennae and mouthparts, the former being concerned mainly with smell, the latter with contact chemoreception ('taste'). The antenna of fifth instar *Schistocerca* consists of 23 annuli with a total of 680 olfactory receptors which have been shown to respond to a very wide range of chemicals. In addition, there are 400 coeloconic sensilla each of which is sunk into a pit below the general surface of the cuticle. These sensilla have been shown to respond to water vapour. Most sensilla are concentrated on the distal annuli of the antenna, with far fewer in the proximal region.

Very large numbers of sensilla are present on the mouthparts—a total of 15 000 in adult *Locusta* and 12 500 in adult *Schistocerca*. Many of these are mechanoreceptors, that is sensilla which respond to mechanical stimuli and which are concerned with controlling the movement of the mouthparts and monitoring the physical presence and movement of the food. A few are olfactory receptors, but a very large number are contact chemoreceptors. The more important of these are in groups, which are shown in Fig. 3–1. Some of these have been examined electrophysiologically and found to respond to a wide range of chemicals. Those on the tips of the palps, for instance, respond to inorganic salts, sugars, amino acids and other chemicals, and less extensive studies in the A_1, A_2 and A_3 sensilla (Fig. 3–1) indicate a comparable extensive range of sensitivities.

Unlike the situation in the contact chemoreceptors of the fly, the neurons in these receptors seem to be relatively unspecialized in the chemicals they respond to. There does not, for instance, appear to be a specific sugar-sensitive cell, though some cells may respond more vigorously to sugars than to other substances, and it is likely that discrimination of different chemicals occurs within the central nervous

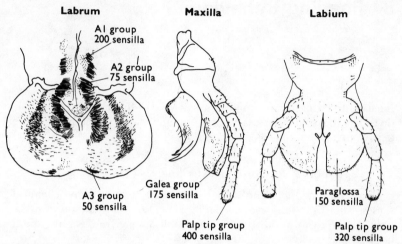

Fig. 3–1   Diagrams of the epipharyngeal surface of the labrum, a maxilla and the labium of a locust. The position of each sensillum is shown and the main groups of chemoreceptors are labelled, with an indication of the number of sensilla in each group. This number varies with the size and species of locust, but the positions are constant. Most of the long hairs on the labrum are non-sensory.

system rather than at the level of the receptors. This has also been found to be true in the discrimination of odours.

For details of the structure of insect chemoreceptors see LEWIS, 1970. For general physiology of insect senses see DETHIER, 1963. For studies on discrimination of chemicals by locusts see BLANEY, 1974.

3.2   Food finding

Before a locust starts to eat it has first to find a potential food plant and then to establish that it is suitable food. These two aspects are differentiated as food finding and food recognition. At least under some circumstances the problem of food finding does not occur, for the insect is more or less constantly on its food plant. This is true of *Nomadacris* in its outbreak areas (p. 48) where it is almost invariably found on some grass which is suitable for food. With other species, however, living in more sparsely vegetated and very arid regions, locating food from a distance may be important. This probably applies to *Schistocerca* living in a desert habitat and to *Chortoicetes* sometimes occurring in very arid inland areas of Australia.

Where plants occur sparsely in an otherwise open area it is very likely that the insects are initially attracted by the visual stimulus presented by the plant. Experiments have shown that a pattern of vertical stripes is especially attractive. In addition it seems clear that olfactory stimuli are also important. It is recorded that large numbers of *Dociostaurus* flew to a patch of moistened soil and it is conceivable that partly desiccated insects respond to moist air by turning and moving upwind so that they are brought to the vicinity of the moisture source. Food odour may be important if the insects have been without food for some time and in the laboratory *Schistocerca* nymphs deprived of food for four or five hours respond to the odour of food by moving upwind. The suitability of the plant as food is determined by a new set of stimuli.

3.3   Food selection

Some locusts are far more discriminating in their choice of foods than is commonly imagined. *Locusta, Locustana, Chortoicetes* and *Nomadacris* are all essentially grass feeders; with a few specific exceptions they rarely eat broad-leaved plants. *Schistocerca* on the other hand is much more catholic and eats many different types of plants, including grasses. Nevertheless, it still exhibits preferences and given a choice will take certain plants, such as *Heliotropium*, in preference to others.

In practice, the choice of food is affected by many different factors: ecological, behavioural and physiological. Even the grass-feeding species eat a sufficiently wide range of plant species not to be restricted in their general distribution by availability of food; rather the converse is true—the insects eat what is available to them in the habitat. *Nomadacris*, for instance, eats the sedge *Cyperus longus* at times when this plant is common in the habitat, but later in its development, when the *Cyperus* dies off, it feeds on grasses such as *Cynodon dactylon*. Similarly where the grass *Vetiveria* is common it forms an important part of the diet, but there are many regions in which *Nomadacris* survives and develops well in the absence of this grass. With migrant swarms the food eaten must clearly depend on where the locusts happen to land.

Feeding cannot be considered in isolation from other activities since these will affect the range of food available to the insect. Migration influences choice in this way, but so do less conspicuous facets of behaviour. For instance, *Nomadacris* makes daily movements between tall and short grass, moving into tall grass where it roosts at night and down into short grass during the day. Consequently, the last meal at night and the first in the morning, when the locusts are warming up, tend to be on the taller grasses, while during the day short grasses provide the main food. As this statement implies, feeding activity, like other aspects of behaviour, is inhibited by low temperatures, and commonly very little food is ingested when the body temperature is below 20°C.

Within these broad limits set by the ecological and behavioural features of the insects' biology, what is eaten depends on the physical and chemical characteristics of the plants. Very hard plants tend not to be eaten, although *Locusta* nymphs will eat very hard bamboo leaves. In this case they simply take longer to eat a full meal. There is evidence amongst grasshoppers that very hairy leaves may be protected from small nymphs which find it difficult to bite the leaf surface. The protection afforded to the plant by these features depends on the extent to which the insect has previously been deprived of food. Recently fed insects will attempt to bite at the leaf, but may be sufficiently deterred by the physical difficulties to move off the plant altogether. Insects deprived of food for longer periods, however, are more persistent and after repeated attempts may finally eat the leaf.

Whether or not a plant is eaten depends ultimately on its chemical characteristics and especially on the occurrence of feeding deterrents. Both *Locusta* and *Schistocerca* are stimulated to eat inert substrates, such as elder pith or filter paper, by a wide range of sugars which are common in most plants. Some common amino acids are also stimulatory. So it seems very unlikely that selection depends on the presence of special stimulating chemicals because suitable substances are found in all plants. On the other hand it has been found that feeding on an acceptable substrate is limited or entirely inhibited by a wide range of chemicals found in plants or by extracts of the plants. In the case of *Locusta*, which feeds almost entirely on grasses (Fig. 3–2), all the plants which it does not eat contain such deterrent chemicals.

Many different classes of chemicals are involved: alkaloids, glycosides, terpenoids and tannins, for instance, and the effect depends on their concentration. It does not follow that, for instance, all alkaloids inhibit feeding in concentrations similar to those occurring in plants. Some grasses are eaten although they contain alkaloids, but other plants may be avoided because of the alkaloids they contain. The same applies with other classes of compound and even inorganic salts may inhibit feeding if they are present in high enough concentrations.

Schistocerca is totally inhibited from feeding by only a few plant extracts in relatively high concentrations, reflecting the very broad spectrum of plants which it eats (Fig. 3–2). One notable exception to this generalization is the extract of *Azadirachta indica*, the neem tree, which contains the triterpenoid azadirachtin. This tree appears to be almost totally protected against *Schistocera* by this chemical and the extract of the seeds has been applied to other plants in India for many years to protect them from locust attack. In the laboratory, *Schistocerca* is deterred from eating sucrose impregnated filter paper by a $10^{-6}\%$ solution of azadirichtin. *Locusta* is much less sensitive to the chemical.

For further information on factors affecting food selection see CHAPMAN, 1974.

3.4   Amounts of food eaten

A critical factor in the importance of locusts is the amount of food they

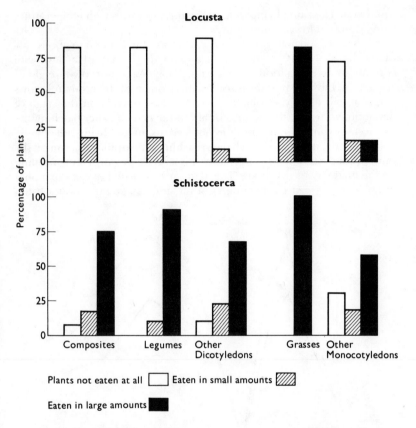

Fig. 3-2 The food preferences of *Locusta migratoria* and *Schistocerca gregaria*, showing the percentage of plants in each plant family or group which were not eaten at all, eaten in small amounts or eaten in large amounts. *Locusta* feeds mainly on grasses, rejecting the other plants; *Schistocerca* eats a wide range of plants.

ingest. This will obviously depend on the suitability of the food, but assessments are normally based on the assumption that the insects are eating entirely acceptable food.

Isolated locusts in the laboratory tend to take food in discrete meals lasting several minutes with intervals of an hour or more between; they do not usually nibble continuously, although very active insects may tend to do this. The amount eaten in one meal depends on the time since the last meal because, if undisturbed, a locust will feed until its foregut is full. After a meal the food starts to pass back to the midgut and in six to seven hours at 30°C the foregut is completely emptied. So the longer the interval between meals the emptier will the foregut be and so the more the insect can ingest. The rate of emptying of the foregut varies with temperature, being slower at lower temperatures.

A full meal, starting with the foregut empty may amount to about 15% of the body weight of the insect, and in the course of a day, with continuous access to food, the locust may ingest about its own weight of vegetation. Obviously as the insect moults from instar to instar and grows bigger it will eat more, but the amount also depends on the stage of development within the instar. For about the first day after hatching the first instar nymph commonly does not feed, its midgut being full of yolk. Subsequently, it eats more each day, reaching a maximum amount on the third or fourth day of the instar. After this it eats less again, eating nothing on the final day of the instar. This pattern is repeated in each nymphal instar (Fig. 3–3). At the beginning of the adult stage the amount eaten

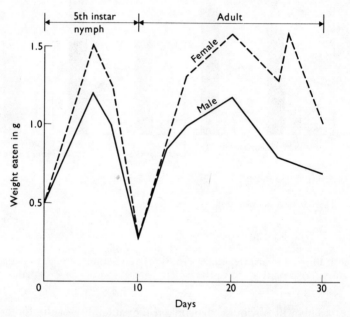

Fig. 3–3 Average weight of fresh grass eaten per day by male and female *Schistocerca* during the fifth nymphal instar and early adult stage. (After DAVEY, P. M., 1954, *Bull. ent. Res.*, **45**, 539–51.)

increases to a high level during the period of somatic growth, for at the time of adult emergence, development of the cuticle and the flight muscles is not complete. Subsequently less food is eaten until, in the female, egg development begins; at this time food intake is again increased to provide the materials necessary for egg production (Fig. 3–3).

3.5 Damage caused by locusts

All these factors will affect the damage caused to crops by locusts. Locusts do not seek out crops to attack, but because crops are grown in relatively well-watered areas they are often greener and more lush than the surrounding bush and so the locusts may tend to feed more actively on them. The damage done will depend on the amount eaten by individual locusts, by the density of locusts and by the time they remain in an area.

The amount eaten by an individual varies with its state of development (Fig. 3–3) and this is sometimes reflected in the damage done. Fig. 3–4 shows the monthly distribution of reports of damage done by the red

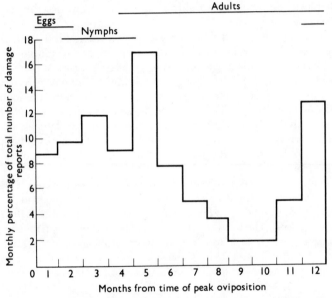

Fig. 3–4 Monthly distribution of reports of damage by the red locust in east Africa from 1930 to 1940 in relation to the stage of development of the insects. (After BULLEN, F. T., 1966, *J. appl. Ecol.*, **3**, 147–68.)

locust in Kenya, Uganda and Tanganyika from 1930 to 1940 in relation to the life cycle of the insect. To judge from the abundance of reports most damage was done by the adults soon after moulting and at the time of egg production. This would be expected from the individual variations in feeding.

A single female *Schistocerca* may eat 1.5 g of vegetation per day; extensive damage only occurs when the locusts are in high density. Adult densities in settled swarms range from about 30 to 150/m² so that amounts eaten could vary from 45 to 225 g/m^2/day if the insects remained feeding in the area all day. In a swarm covering an area of 10 km^2 this would amount to a total of between 0.5×10^6 kg and 2.25×10^6 kg (about 2000 tons) of vegetation eaten in one day.

In terms of economic loss the effect of locust damage to a crop will depend on the stage at which the plant is attacked. The seedling is very vulnerable because it may be completely destroyed. In older plants, defoliation may only cause a temporary delay in development because many plants, like maize, can regenerate and produce a full yield. With maize a second vulnerable stage is at the time of flowering and if the female inflorescence is injured the crop may be almost totally lost. The maize cobs may also be attacked, and damage at this stage is important because the grains cannot regenerate, but as the cobs get older they are less liable to attack. With sorghum the unripe grain may be eaten in preference to the foliage.

As would be expected from its wide range of natural host plants, *Schistocerca* attacks a wide range of crops; some of the most important are wheat, sorghum, millet, rice, pigeon pea and citrus. *Locusta* is essentially graminivorous and the crops most commonly and extensively attacked are maize, sugar-cane and sorghum. Nevertheless there are instances of *Locusta* damaging bananas, ground nuts and other broad-leaved plants. With insects in such vast numbers extensive damage may be inflicted even if they take very small meals, or if, as sometimes happens, they feed on the stems and leaf petioles which appear to be less distasteful than the leaves.

Damage to pasture may also be severe. There are no known figures directly concerning this, but a cow eats about 12 kg of vegetation per day, and the grazing capacity of land in the tropics is often of the order of 15 animals per km^2. The food consumption of locusts in swarm density over the same area might be as much as 150 000 kg—about 1000 times as much as the cattle would eat. Obviously swarms may have a disastrous effect on the grazing capacity and it may take years before such pasture recovers.

4 Flight

The flight of locusts has been more fully studied than that of any other insect, partly because of their economic importance, but also because of their large size and ease of handling in the laboratory.

4.1 Flight mechanism

The wings are formed of a double layer of cuticle and it is important to remember that, as in all other insects, there is no break in the cuticle where they articulate with the thorax. At this point the cuticle is flexible and membranous except for four small axillary sclerites which connect the thorax with the bases of the wing veins. Beneath the wing is the main hinge, which consists of a pad of a rubber-like cuticle, called resilin. In the relaxed position this resilin pad holds the wing down. When the wing is elevated by the muscles or by hand, the resilin is stretched and much of the energy used in elevating the wing is stored in the pad and supplements the depressor muscles in pulling the wing down again.

The wings are moved by distortions of the thorax. The depression of the dorsal surface by dorso-ventral muscles causes the wings to move up because of the nature of their articulation. Bowing upwards of the dorsal surface, resulting from the action of dorsal longitudinal muscles, causes the wings to move down. The muscles producing these effects are known as the *indirect flight muscles* because they move the wings indirectly as a result of distortions of the thorax. The *direct flight muscles*, on the other hand, are inserted directly into the bases of the wings. They assist in wing depression, but are also involved in the control of flight by producing twisting movements of the forewings.

For details of the anatomy of the flight mechanism see ALBRECHT, 1956.

4.2 Aerodynamics of flight

The insect is kept in the air and propelled along by the flapping of the wings. The forces produced by these movements are complex because the movements of the wings themselves are complex: the wing tip follows an irregular path as the insect flies along and at the same time the forewing is twisted by the indirect muscles so that its leading edge is sometimes higher, sometimes lower than the trailing edge (Fig. 4–1). Twisting of the hindwing is passive, the large vannal area being forced into different positions by air pressure.

As a result of these movements the insect always produces positive lift forces, even during the upstroke (Fig. 4–1), and although the thrust force is more variable the overall effect is to produce continuous forward movement. About 90% of the energy expended is involved in keeping the

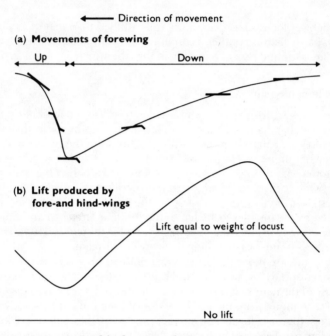

Fig. 4–1   (a) Movement of the forewing of *Schistocerca*. The continuous line shows the path of the wing tip as the insect flies forwards, the thick lines show the angle at which the mid point of the wing is held at different phases of the stroke. (b) The lift force produced by both pairs of wings. At no time is a negative lift, that is a force pushing the insect down, produced although during the upstroke of the wings the lift force is less than the downward force produced by the weight of the locust. (After WEIS-FOGH, T., 1956, *Phil. Trans.*, B **239**, 459–510.)

insect in the air (lift), and some 70% of all the force is exerted by the hindwing.

4.3   Control of the wingbeat

The first beats of the wings are stimulated by loss of tarsal contact with the ground as the insect leaps into the air. Subsequently the role of driving the wings is taken over by the aerodynamic sense organs on the head. These consist of groups of small hairs (Fig. 4–2) which are bent by the flow of air over the head as the insect moves forward; bending provides the stimulus for continued beating of the wings. Their role can be demon-srated by blowing on the face of a locust suspended on the end of a nail or pencil.

The rate at which the wings beat is relatively constant at 20–25 Hz in a

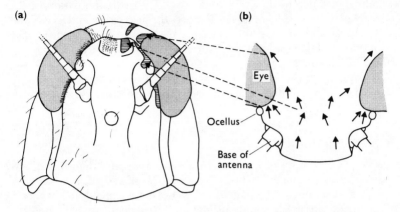

(a)

(b)

Fig. 4–2 (a) Front view of the head of *Schistocerca* showing the positions of the aerodynamic hairs. (After WEIS-FOGH, T., 1949, *Nature, Lond.,* **164,** 873.) (b) View of top of the head showing the directions in which the hairs are most sensitive to bending. Hairs in equivalent positions in the two diagrams are indicated by broken lines. (After CAMHI, J. M., 1969, *J. exp. Biol.,* **50,** 335–48.)

fully developed locust and this is controlled by a rhythmic activity within the central nervous system which causes the muscles to contract at the appropriate times. Sensilla, which function as proprioceptors, at the base of the wing feed back information to the central nervous system on the precise position and movement of each wing and the information they convey provides a fine adjustment to the basic control rhythm. In this way uncontrolled movements of the body which may be imposed by turbulent air or loss of stability by the locust are immediately corrected for.

Lack of uniformity of the wing beat or uneven air pressures may cause the insect to move from its straight path in the vertical or horizontal planes. Such deviations are called roll, if the insect rotates about its longitudinal axis, pitch, if the head moves up or down relative to the abdomen, or yaw if the head tends to waver sideways relative to the abdomen. The insect is able to perceive these deviations by various sense organs and so can adjust its wing beat to correct for them.

Roll is apparently controlled visually. Locusts, like many other insects, have a dorsal light reaction: they tend to orientate so that the dorsal ommatidia of the compound eyes are more strongly illuminated than the ventral ones. In addition, the locust stabilizes its position in flight by keeping the horizon level across the two eyes. If the insect deviates from the situation where the maximum illumination is from above and the horizon is level across the eyes, differential changes in the wingbeat of the two sides are made to restore the position.

Deviation from a straight path resulting in yaw is perceived by the aerodynamic sense organs on the head, which respond differentially to bending in different directions (Fig. 4–2). Unequal stimulation of the sensilla on the two sides of the head causes the insect to extend its legs and twist the abdomen so as to steer itself back on to its original course. Movements of the head relative to the rest of the body also evoke these leg and abdominal movements.

For a general account of the physiology of insect flight see PRINGLE, 1965. For details of flight control mechanisms see BURROWS, 1975; CAMHI, 1970.

4.4   Energy for flight

The power output of insect flight muscles is greater than that of any muscles in any other animal; the flight muscles of a locust produce over 35 W/kg when the insect is in flight. This enormous expenditure of energy depends on an ample supply of fuel and a mechanism facilitating the rapid oxidation of the fuel with the release of energy.

The two main fuels available to a locust are carbohydrates and fats. These are normally present in the haemolymph, but are also stored in the fat body. The muscles are directly bathed in haemolymph, since the locust like all insects has an open circulatory system, and the haemolymph penetrates between the muscle fibres and into fine channels invaginating into them. But to maintain an adequate supply of fuel at the mitochondria, where oxidation occurs, very high concentrations are necessary in the haemolymph. Carbohydrate levels fall at the beginning of flight (Fig. 4–3) and are not replenished, so it is likely that carbohydrates provide the main fuel for flight only for a short time after take-off. The concentration of lipids in the haemolymph, on the other hand, starts to rise after about 10 minutes flight and is then maintained at a high level (Fig. 4–3). These lipids are taken from the fat body where they are stored, the process being controlled by an adipokinetic hormone from the glandular lobes of the corpora cardiaca. The release of the hormone is prevented if nerves to the brain are cut and the locust is then unable to make sustained flights.

Fats have some advantage over carbohydrates for migrant animals. Twice as much energy is produced from the oxidation of a unit weight of fat as from the same weight of carbohydrate, and at the same time approximately twice as much metabolic water is produced. This could be critical in an insect flying for a long period since some of its spiracles are continuously open and water loss may be considerable.

The supply of oxygen to the flight muscles is assured by the anatomy of the thoracic tracheal system together with a modification of the resting

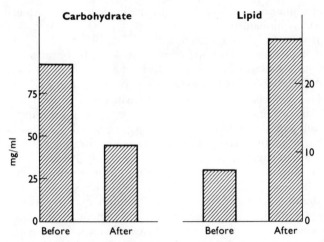

Fig. 4–3   The carbohydrate and lipid concentrations in the haemolymph of *Locusta* before flight and after 30 minutes flight. (After Wajc, E., unpublished.)

pattern of opening and closing of the spiracles. The main air supply to the flight muscles runs directly from the first thoracic spiracle and is largely isolated from the rest of the tracheal system. Within the flight muscles the tracheae expand into extensive airsacs from which fine tracheae indent the muscle membrane and approach close to the mitochondria.

In the resting insect, ventilatory movements of the abdomen together with the opening and closing of the spiracles maintain a flow of air through the tracheal system from front to back. During flight, however, the first spiracle, which leads to the flight muscles, remains permanently open. The contractions of the flight muscles produce alternate collapse and expansion of the airsacs within so that air is alternately forced out from and drawn in through the permanently open spiracle—a tidal flow. In this way an ample supply of oxygen is maintained and the carbon dioxide produced in the course of oxidation is pumped directly to the outside rather than round the rest of the body.

For further information on the power output of locust flight muscles see WEIS-FOGH, 1964.

4.5 Flight in relation to the physiology of the insect

When the locust moults from the nymphal form to the adult it has fully developed wings, but it is at first unable to fly. Over the first few days of adult life the cuticle hardens and thickens and the flight muscles become fully functional. During this period the insect may take off but the flight is of short duration. This period during which sustained flight cannot occur is called the teneral period. The insect retains its capacity to fly for the rest

of its life, but the most extensive flights are made by locusts while they are
still sexually immature.

Even when the locust is fully capable of flying other factors may limit or
prevent flight activity and body temperature is one of the most important.
The flight muscles only function efficiently when their temperature is
above 30°C. At lower temperatures the enzymes regulating energy release
are relatively inefficient and the muscle twitch is of long duration. As a
result, the activities of antagonistic muscles pulling the wings up and
down will tend to overlap and much energy is wasted; it is only when the
muscle temperature exceeds 30°C that the twitch duration is short enough
for this not to happen.

Flight itself raises the body temperature, due to the production of
metabolic heat, by about 8°C, so that sustained flight may occur even
when the air temperature is below 30°C. At air temperatures above 40°C,
however, sustained flights do not occur. The physiological reason for this
is unknown, but the adaptive significance is obvious. If locusts made long
flights at air temperatures of 40°C their body temperatures would
approach 50°C, very close to their upper lethal temperature, and flight
would be a hazard to survival.

Provided the body temperature is maintained within the limits of about
30 to 38°C, sustained flight is possible. Fig. 4–4 illustrates the thermal

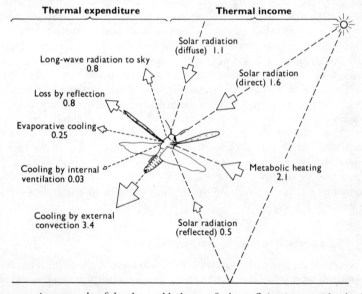

Fig. 4–4   An example of the thermal balance of a large flying insect with a body
temperature about 10°C above air temperature. The numbers indicate cal/min
and are based on observations made on the red locust. (After RAINEY, R. C., 1974, *A.
Rev. Ent.*, **19**, 407–39.)

balance of a flying locust assumed to have a body temperature about 10°C above air temperature. As body temperature is greatly influenced by the air temperature and by radiation from the sun both these factors will profoundly affect the tendency to make sustained flights (Fig. 4–5).

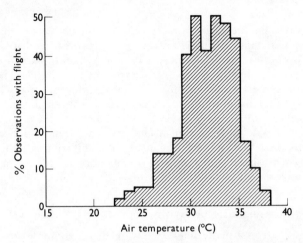

Fig. 4–5   Flight activity of the red locust in relation to air temperature. (After CHAPMAN, R. F., 1959, *Behaviour*, 4, 300–34.)

4.6   Environmental factors affecting flight

Apart from factors directly affecting air temperature, flight may be limited or inhibited by other environmental factors. There is some evidence, for instance, that flight is more sustained at high relative humidities and high winds may inhibit take-off or cause low-flying locusts to land.

It must not be supposed that flight invariably occurs if the environmental factors are suitable. Flight is only one aspect of the behaviour of the locust and it is integrated with other activities such as feeding and reproduction, but at times flight appears to be the dominant activity. Under these circumstances migration occurs.

For an outline of the factors regulating flight activity see HASKELL, 1966.

5 Migration

Migration in locusts is usually taken to imply a mass (swarm) flight away from the immediate breeding area, commonly involving displacements over many hundreds of kilometres. This is only a part of the story. Swarm displacement is determined to a large extent by the winds (see below) so that there may be occasions on which a swarm flies for many hours or even days with virtually no displacement at all. The insect is behaving in the same way whether it is displaced a long way, or not at all, and it is more useful to define migration in a behavioural/physiological sense as a period in which flight is the dominant activity. This contrasts with 'trivial' flights which are an integral part of the normal feeding or mating behaviour. Solitary locusts may also migrate.

5.1 Daytime migration by swarms

Locust swarms usually migrate during the daytime. Most work on swarming flight has been on *S. gregaria* and this account is based largely on this species, but it may be supposed that the same principles apply to the other locust species even though the details vary.

In order for a swarm to persist as an entity the individual locusts must react to each other; if they did not their activities and air turbulence would soon cause them to disperse and the swarm would cease to exist. What holds the locusts together? Photographic analysis of the displacements of individuals in a swarm shows that, although a majority of the insects flying above 100 m are orientated in the direction of swarm movement, those lower down within the swarm are orientated in a variety of directions, more or less at random (Fig. 5–1). At the edges of the swarm, however, it is observed that all the locusts are heading back into the mass of locusts. It is very likely that the loss of visual contact experienced by locusts flying out of a swarm causes them to turn and fly back in again, but critical experiments have not been performed.

Within a swarm it is important that the locusts are spaced out so that they are not continually in collision. The average distance between locusts in swarms varies from about 1 to 12 m, but in over 50% of the observations made the locusts are 2–4 m apart. This suggests that they have a 'preferred density' at which they fly, indicating not only that there are 'attractive forces' keeping them together in a swarm but also 'repellent forces' preventing them from getting too close. Vision could again be important here, but so could the noise of other locusts in flight and the turbulence of the air created by them.

33

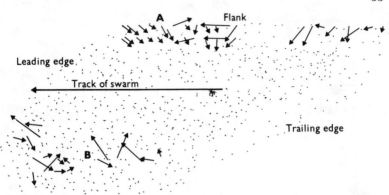

Fig. 5–1 Orientation of locusts flying at low levels in a swarm. Each arrow indicates the mean orientation of a group of locusts, the length of arrow being proportional to the number of insects, while the dotted area indicates the general mass of locusts in the swarm. Notice that at the edges of the swarm [as at (A)] the insects tend to be flying towards the mass of locusts whereas within the swarm [(B) at the lower edge of the illustration] they are orientated in a variety of directions. (After WALOFF, 1972.)

The height at which locusts fly in a swarm varies from just above the level of the vegetation to over 1000 m above the ground. On hot days the sun warms the air close to the ground causing it to become less dense than the cooler air above. The hotter air rises and the lift normally imparted to the locust by the movement of its wings is supplemented by the upward movement of the air so the insect will tend to rise rapidly. Since the hot air is replaced near the ground by cooler air from above there will also be airflows tending to carry the insects down. The locusts are not entirely at the mercy of these air movements, but are greatly affected by them. As a result, on hot days when there are strong thermals the locusts will extend from ground level to a considerable height, perhaps 1000 m or more above the ground. These towering swarms are called cumuliform swarms by analogy with cumulus clouds and the upper limit of the swarm coincides with the limit of thermal convection (Fig. 5–2). At times when there is little or no convection the insects tend to remain close to the ground and under these conditions a stratiform swarm is found.

Locusts can fly at airspeeds of 10–25 km/h so that as long as windspeeds are lower than this the locusts are capable of moving in any direction relative to the ground. At higher windspeeds they will be displaced downwind whichever way they are orientated. It is very common to see locusts flying into the wind at very low windspeeds and they may move a few miles upwind, but the long-range migrations are the result of carriage on the wind. This arises because locusts in big swarms continue to fly in winds which exceed their individual air speeds, and because the locusts

34

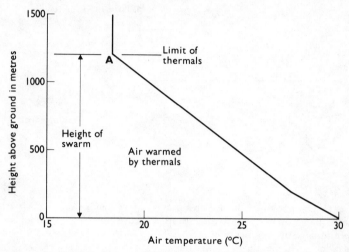

Fig. 5-2 The height of a cumuliform swarm of *Schistocerca* in relation to air temperature. Above point (A) the temperature is more or less constant; below it the air is warmed by thermals. The upper limit of the swarm roughly coincides with the upper limit of the thermals. (After RAINEY, R. C., 1958, *Q. Jl R. met. Soc.,* **84,** 334–54.)

within a swarm are randomly orientated with respect to the wind and the ground. The speed of displacement downwind is usually less than the windspeed, but it is relatively common for a swarm to be displaced 50 km in a day and distances of over 100 km/day are recorded. There are occasional records of much longer flights and one of the best documented is of a flight of desert locusts from southern Morocco to Portugal, a distance of at least 1000 km, in less than 24 h in October 1945. It is known that at 500–600 m above sea level a 50 km/h wind was blowing from the south and it is almost certain that the locusts were carried up to this height by strong thermals and then swept northwards.

This tendency for swarms to move with the wind is important in two ways: it has survival value for the locusts and it produces a fairly regular, and so predictable, pattern of migratory movements. It is of survival value because the winds are moving into zones of convergence, that is regions in which the net horizontal inflow of air exceeds the outflow. These are regions in which rain tends to fall so that locusts are carried to areas where there is likely to be adequate moisture in the ground for egg development and where fresh young plants will be growing for the newly emerged nymphs to feed on. In a species like the desert locust which traverses extensive areas of arid land this is clearly of critical importance.

The dominant influence in the distribution area of the desert locust is the Intertropical Convergence Zone (ITCZ) which marks the boundary between dry continental air from the northern hemisphere and moisture-laden equatorial air, but there are also other important convergence systems. Because these systems develop in a regular seasonal manner the major migration patterns and breeding areas of the desert locust are also seasonal. In the summer the ITCZ extends from West Africa to India and summer breeding can occur throughout this vast area (Fig. 5–3). During

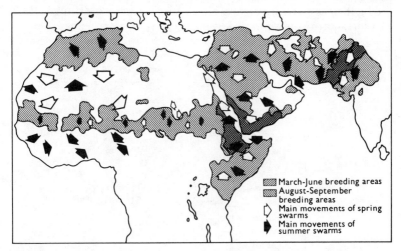

Fig. 5–3   The main breeding areas and migratory movements of *Schistocerca* at different times of year. The winter breeding areas in eastern Africa have been omitted. (After WALOFF, 1966).

the autumn the locusts produced in West Africa move northwards to north-west Africa, but they do not breed until temperatures rise in the spring. The progeny of this spring breeding then returns across the Sahara to be concentrated along and breed on the ITCZ. On a second migration circuit, the offspring of swarms breeding on the ITCZ in southern Arabia, eastern Africa and the Sudan migrate into the trench formed by the mountains on either side of the Red Sea and the Gulf of Aden to breed during the winter and early spring rains associated with the Red Sea Convergence Zone. The adults produced there move north towards the Middle East and northern Arabia and breed again in the spring before their offspring return south to the area of the ITCZ. On this circuit three generations are produced each year. Other migration circuits of the desert locust exist in eastern Africa and between the Middle East and India.

In many years these migration circuits are not self-contained and some of the most extensive migrations occur when swarms move from one circuit to another. More local movements are superimposed on this general pattern so that the distribution of locusts at any one time is more extensive and complex than this simple account would suggest.

The migrations of the desert locust thus involve movements back and forth between successive breeding grounds, though the movements may be irregular and involve individuals of successive generations. A single individual does not make the outward and return flight which is regarded as a characteristic of bird migration. In other species, such as the red locust, there is no return flight at all, the displacements constituting more of a spread from one or more foci (Fig. 5–4).

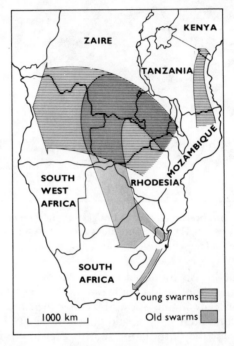

Fig. 5–4   The major annual displacements of swarms of red locusts during the 1933–43 plague. (After SYMMONS, P. M., unpublished.)

There is no reason to think that the differences between these species are fundamental. Sustained flight only occurs when the insects are in a particular physiological state and when environmental conditions are suitable. The direction of displacement depends on the winds. If there are regular seasonal reversals of the wind which coincide with other conditions permitting flight, then a two-way migration will occur. If, however, the flight period is limited to times when the wind is predominantly in one direction, then expansive migration will occur.

For full accounts of the migrations of the desert locust see RAINEY, 1963;

WALOFF, 1966. An important paper on the orientation of locusts in swarms is by WALOFF, 1972.

5.2   Migration by solitary locusts

Locust swarms characteristically migrate during the daytime, but solitary locusts seem only to make long range flights at night. In the case of *S. gregaria* and *Chortoicetes* take-off is stimulated by rapidly falling light intensity 15–40 minutes after sunset. Sustained flight only occurs on warm nights since, in the absence of the sun, the body temperature of the locust will depend primarily on the air temperature and its own output of metabolic heat. If the air temperature is too low the rate of heat loss will exceed heat production by the insects so that the body temperature will fall, the muscles will cease to function efficiently and the insect will land. The evidence so far accumulated suggests that extensive night flight does not occur when the air temperature is below 22°C.

Using radar it has been shown that solitary desert locusts fly as high as 1.8 km above the ground, although the majority of insects are at lower levels, below 400 m. These insects are displaced downwind and, it appears, are also commonly orientated downwind. Their rate of movement obviously depends on the strength of the wind, and in one series of observations in the Niger Republic the average speed was 35 km/h, with some insects travelling at 65 km/h.

These flights may result in extensive displacements of solitary locust populations. The fullest studies have been made on the migratory locust in and around the flood plains of the River Niger in Mali. Many thousands of locusts were marked with paint and marked individuals were recorded as much as 250 km away from the point of marking (Fig. 5–5). The effect of the migrations is first to carry the insects away from the flood plain into the adjoining semi-arid country. Here the insects breed wherever seasonal rainfall produces vegetation and soil suitable for oviposition. The adults of the next generation return to the flood plains as the floods recede and the semi-arid areas, now rapidly drying out, are completely evacuated.

Comparable night flights by *S. gregaria* have been observed in several parts of its range. For instance a regular movement involving displacements of up to 1000 km is observed between the winter and spring breeding areas in the extreme west of Pakistan and the summer breeding areas in eastern Pakistan and western India. A return movement, comparable with that made by swarms, is also known to occur.

For an account of the night flight of desert locusts see ROFFEY, 1963. For the use of radar in tracking locusts see SCHAEFER, 1975.

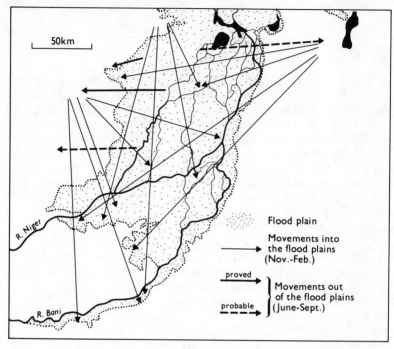

Fig. 5–5 Flights by solitary *Locusta* to and from the flood plains of the River Niger as indicated by the recovery of marked individuals. Only a few representative examples of the recoveries made are illustrated. (After DAVEY, J. T., 1959, *Locusta*, **7**, 1–180.)

5.3 Migration by nymphs

Locust nymphs, like the adults, are gregarious and they group together to form bands (the equivalent of adult swarms). Bands vary in size from a few square metres up to several square kilometres. The nymphs in a band migrate (march) by walking or by making frequent short hops of a few centimetres (quite different from the much longer escape hops).

The distance covered by nymphal bands depends on a number of factors. Big bands move further than small ones, presumably because there are more interactions between individuals in a bigger band. Distances moved are, of course, much less than those covered by adult swarms and it is recorded that a small band, comprising 10 000–20 000 locusts, in east Africa moved about 3 km during the nymphal life of the insects, while a band which initially contained over 5 000 000 nymphs moved about 20 km. Movement is also reduced before moulting and by dense vegetation. Sparse vegetation provides less mechanical obstruction to the insects, enables them to see and interact with each other more readily, and provides fewer opportunities for feeding.

The direction of movement is affected by numerous factors. The wind is particularly important: most bands and especially small ones tending to move downwind. Changes in vegetation and open spaces may affect the line of movement by providing barriers or by canalizing the insects. Once the direction of movement is altered, the new direction tends to be maintained for the remainder of the day.

It is not clear whether or not movements by bands of nymphs should be regarded as strictly comparable with adult migration. The movement does not carry the insects outside the normal breeding area, nor does the marching appear to be a dominant activity in the same way as flight, because the insects stop to feed frequently and in small bands individuals can be observed to undergo cycles of feeding, resting and marching just as isolated locusts have cycles of feeding, resting and moving. Perhaps this 'migration' is to be regarded more as an exaggeration of the normal bouts of activity which occur in a normal feeding cycle and which tend to ensure that the insect is presented with a variety of food plants to eat. The exaggeration could result simply from gregarious interactions between insects tending to make them more active.

For further details see ELLIS and ASHALL, 1957.

5.4   The significance of locust migration

Locusts do not start to migrate when, or because, they have exhausted their food supply, though we can suppose that the migratory habit arose as a result of pressures imposed by food shortages. The most reasonable assumption is that migration is an adaptation which enables the locust to exist in regions where the food supply is very seasonal and perhaps irregular. It is an obvious advantage for the insect to move before the habitat becomes untenable, and an even greater advantage if it then behaves in a manner most likely to bring it to a new breeding ground. The tendency to move downwind into areas of convergent winds has just this effect.

For general accounts and reviews of insect migration see SOUTHWOOD, 1962; JOHNSON, 1969.

6 The Phases of Locusts

6.1 Introduction

Locusts are not found perpetually in swarms. There are periods, sometimes of many years, when no swarms of a particular species are recorded (see Fig. 7–1). Where do they go during these periods? This remained obscure until, in 1921, B. P. Uvarov published a taxonomic paper on *Locusta* in which he suggested that what had until then been recognized as two species were, in fact, different phases of the same species. One, the gregarious phase, was the well-known locust; the other was the form in which the locust survived during non-swarming periods, called the solitarious phase. It is now recognized that all the locust species exist in these two phases as well as in a series of intermediate forms. In some cases, as in *Schistocerca* and *Locusta* there are distinct morphological differences between the extreme phases. In other cases the differences are in the overall size of the insect and the relative sizes of different parts of the body, such as the wings and hind femora, and in behaviour.

6.2 Characteristics of phases

In general the solitarious nymphs of all the locust species are characterized by a uniform coloration—most often green or straw coloured depending on the humidity and the colour of the background on which they were reared. Gregarious nymphs on the other hand are distinguished by a striking pattern of black markings on a background of yellow, in the case of *Schistocerca* and *Nomadacris*, or orange, in the case of *Locusta*. In the adults the sharp distinction in colour between solitarious and gregarious forms is lost, although solitarious adult *Locusta* do tend to remain green.

We can only guess at the significance of these differences in colour. It is probable that the tendency for solitarious nymphs to have a similar colour to the background gives them some protection against predators, but no critical experiments have been carried out. The coloration of gregarious nymphs, on the other hand, makes them conspicuous, especially when they are in bands. In some other insects conspicuous coloration and grouping are coupled with distastefulness, but there is no evidence that locusts are distasteful to predators. An alternative hypothesis is that the conspicuous markings facilitate visual interaction between the individuals in a band and so help to maintain gregarious behaviour and band cohesion.

The solitarious phase of *Locusta*, and to a lesser extent *Schistocerca*, can also be recognized by a high crest on the pronotum; in the gregarious phase the pronotum is saddle-shaped and without a marked crest (Fig. 6–1). The phases also differ in size. In *Schistocerca, Nomadacris* and female

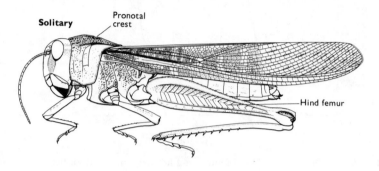

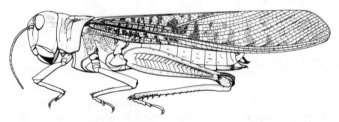

Fig. 6–1 Solitary and gregarious female *Locusta* drawn to the same scale. Notice the greater size, the high pronotal crest and the much larger hind femur of the solitary insect. (From UVAROV, 1966.)

Locusta solitarious insects are bigger than gregarious ones, but in *Locustana, Dociostaurus* and male *Locusta* the converse is true. In *Schistocerca* and *Nomadacris* the bigger size results from the interpolation of an additional nymphal instar into the life history, *Nomadacris*, for instance, having seven or even eight instars in the solitarious phase instead of the six usual in gregarious development. The ratios of various parts of the body to each other are also correlated with phase and one of the most useful is the ratio of head width (commonly called C) to hind femur length (F). This ratio (F/C) is greater in solitarious than in gregarious insects (Fig. 6–2).

The important biological differences between solitarious and gregarious locusts are in their physiology and behaviour. Gregarious insects produce hatchling first instar nymphs which are dark in colour and heavier than those produced by solitarious adults. These hatchlings

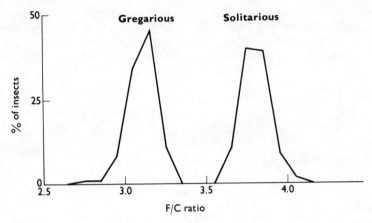

Fig. 6–2 The F/C ratio (hind femur length/maximum width of head) of adult male *Schistocerca* collected from swarms (gregarious) and from low density populations (solitarious).

have a higher water content and contain more fat, haemolymph amino acids and ascorbic acid than their solitarious counterparts. As a result they can live for longer when deprived of food. Solitarious nymphs show little or no tendency to aggregate because this is learned during chance encounters with other locusts. Gregarious nymphs aggregate strongly even in a uniform environment (Fig. 1–1). The latter are also much more active.

An important difference between solitarious and gregarious adults is in the number of eggs they lay. Solitarious females of *Locusta, Schistocerca gregaria* and *Nomadacris* lay more eggs per pod than crowded females, but in *Locustana* and *Dociostaurus* the converse is true. Perhaps this simply reflects differences in the relative sizes of the crowded and isolated females in these species. Gregarious individuals of the first three species also lay fewer pods than isolated females, probably reflecting their shorter life span, so the total number of eggs produced by gregarious insects is considerably lower (Table 3).

6.3 Control of phase

The phase of a locust depends largely on the physical and social environment experienced by the individual insect during its lifespan. This is clearly shown by the fact that solitarious and gregarious insects can be reared from a single egg-pod simply by keeping them in isolation or in a crowd. Not only do the resulting insects differ in appearance, they also differ in their tendencies to aggregate and to march. The importance of individual learning is also indicated by the ease with which solitarious

Table 3   The fecundity and fertility of locusts

Species	Rearing conditions	Average no. of egg-pods /female	Average no. of eggs/ pod	Average total egg pro- duction	Average no. of ovarioles /female	% Resorption
Locusta	isolated lab.	7.0	71	497	103	31
	isolated field	—	79	—	102	23
	crowded lab.	5.6	59	330	83	29
Schistocerca	isolated lab.	5.3	66	350	130	49
	crowded lab.	3.8	69	262	115	40
Nomadacris	isolated lab.	6.0	107	642	—	—
	crowded lab.	4.0	97	388	—	—
	transiens* field	—	128	—	164	22

* Transiens is the name given to forms intermediate between the extreme solitarious and gregarious phases.

individuals can be induced to behave in a gregarious manner. Nymphs of *Schistocerca* which had been reared in isolation up to the fourth instar were made to group together by warming one part of a cage and keeping the rest cool. Even 30 minutes of this 'training' was sufficient to produce a marked increase in the tendency of each individual to join a group of gregarious nymphs, and after four hours' training the solitarious insects grouped almost as strongly as those reared in crowds.

There is, however, also some innate tendency towards one or other of the phases, depending on the treatment of the parents, since the offspring of locusts kept in isolation for several generations show a reduced tendency to march, even when crowded, than do the offspring of crowded parents. The importance of parental treatment is reflected also in the ovariole number and in the F/C ratio, but the treatment of the individual within its own lifetime is still dominant (Fig. 6–3).

An individual insect may be made aware of the presence of others of its own species by physically contacting, seeing, smelling or hearing them. It has been shown that vision is not important in the *development* of gregarious behaviour by rearing isolated insects in cages in which they were able to see other locusts, but were not in contact with them. These insects showed no tendency to group. However, vision is important in the *maintenance* of gregarious behaviour because if gregarious nymphs are tested in the apparatus shown in Fig. 1–1 in complete darkness they show no tendency to aggregate for several hours. The importance of physical contact in developing gregarious behaviour has been shown in a number of experiments in which isolated locusts have been crowded with grasshoppers or even with woodlice or simply touched continually by

44

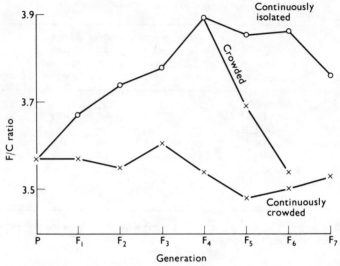

Fig. 6–3 F/C ratios of successive generations of crowded and isolated *Schistocerca* reared from the same parental stock (P). Some offspring from the isolated F_4 generation were subsequently reared in crowds, and it took two generations for the F/C ratio to return to the level of continuously crowded stock. (After HUNTER-JONES, P., unpublished.)

wires dangling into the cage. After any of these treatments the locusts, when tested in groups, aggregated nearly as much as those reared in crowds.

These sensory inputs may modify the behaviour of the insects more or less directly through the central nervous system. More permanent phase differences such as colour and ovariole number entail biochemical modifications which are unlikely to be induced directly via the nervous system. There is some evidence that the endocrine system, especially the corpora allata, is involved. Implantation of corpora allata into gregarious nymphs causes them to become green after the next moult, while their removal from solitary insects is followed by an increase in the amount of black pigmentation. The corpora allata, together with the prothoracic gland, regulate the occurrence of moulting in locusts and so will be involved in determining the number of nymphal instars and the wing length of the adults.

The corpora allata are regulated by neurosecretory cells in the brain and it is via these cells that the information from the various peripheral inputs is finally transmitted to the endocrine system. This may, in turn, lead to feedback to the nervous system itself. As far as is known the sensory systems of solitarious and gregarious locusts are essentially similar and there is no evidence of any difference in the responsiveness of the sense cells to identical visual, acoustic and mechanical stimuli. The differences in response by insects of the two phases must then result from differences within the insects. Very little is known about this, but it is known that haemolymph from crowded *Schistocerca* nymphs injected into solitary nymphs causes them to become more responsive and that an extract from the prothoracic glands reduced marching activity of gregarious nymphs. These examples only indicate that some direct effect of the endocrine system on the nervous system is possible; whether this is normal or how it is produced has yet to be elucidated.

For experimental studies on phase transformation see papers by ELLIS; see ELLIS, 1972, for references.

6.4   Phase transformation in the field

Since solitary locusts show no tendency to aggregate, the first stages of phase transformation in the field must be the result of environmental conditions bringing the insects together. It is possible to envisage a number of ways in which this might happen although the process has never been followed because of the difficulty of observing the behaviour of individual insects for long periods in the field. The bringing together of locusts by factors in the environment is called *concentration*, to differentiate it from *aggregation*, the coming together of locusts due to their gregarious tendencies.

Like many other insects, locusts are able to control their body temperatures to some extent by moving to positions where their bodies are suitably exposed to or sheltered from the sun. This behaviour is most evident in the early mornings when temperatures are relatively low. The insects wander on the ground, or climb the vegetation, until they reach warm spots and here they may sit for long periods. Such warm spots may be provided by stones and small heaps of soil at appropriate angles to the sun, and if these are relatively limited in number and area they may serve as foci for concentrating the locusts. Any feature of the environment which is similarly limited in number and area and in which the locusts tend to remain will serve the same function. Hence patches of green vegetation in an otherwise bare area, or tall clumps of grass rising above the general level of vegetation may serve as 'traps' within which the insects are concentrated.

Once the process of concentration has started it will be reinforced by the insects learning to aggregate. Having once been forced into a group, the locusts will tend to aggregate spontaneously. This is the first step in the process of *gregarisation*, the transformation of a solitarious locust into a gregarious one. If numbers are low they will not develop full gregarious behaviour or colouration, but provided they remain together the process will be continued in the next generation. Hence the development of a fully swarming population from a solitarious population will depend on three interrelated processes: multiplication, concentration and gregarisation.

The converse change from gregarious to solitarious occurs when numbers become reduced so that gregarious interactions are less likely to occur. Environmental factors also play their part here because in a uniform environment there will be little tendency for the insects to come into contact and groups will tend to break up, especially if the vegetation is dense.

It must not be supposed that the change from solitarious to gregarious, or vice versa, is necessarily continuous or direct. The insects may assume a variety of behavioural and morphological characteristics which are intermediate between the extreme phases. They may remain in such an intermediate stage for several generations or may pass directly from one extreme phase to the other. The rate of change will depend on the numbers of insects in the area and on the degree of interaction between them, which will vary with the climatic conditions, the nature of the habitat and the past histories of the insects.

For an account of the build-up of an outbreak see ROFFEY and POPOV, 1968.

6.5   Outbreak areas

Solitarious locusts are widely dispersed through the regions in which the various locusts are recorded as pests. Throughout these regions, and for much of the time, they are no more abundant and no more important economically or ecologically than many grasshoppers. Like the grasshoppers, they undergo fluctuations in abundance, but unlike the grasshoppers they tend to gregarise when their numbers increase. If the numbers are high enough the process will lead to band formation and ultimately to the development of adult swarms.

There is no reason why this should not occur at any point in the range of each species provided the environmental conditions are suitable, and there are records of gregarisation taking place in many parts of the range of *Schistocerca* (Fig. 6–4). In most seasons the breeding success of *Schistocerca* in these places is too low to produce swarms because the rains are too scanty to permit good survival, but in some years sufficient rain falls for the insects to produce a second generation. This results in a sharp increase in population and small swarms may be produced. Such upsurges do not usually give rise to new plagues because the populations are not sufficiently large to permit swarm breeding in the next generation, but on four occasions in the last 50 years, they have led to plagues.

The picture is slightly different for the other two main African locusts—the migratory and red locusts, *Locusta* and *Nomadacris*. *Locusta* is very widely distributed and small swarms have been produced along the flood plains of the middle Niger, around Lake Chad, in the Sudan and in

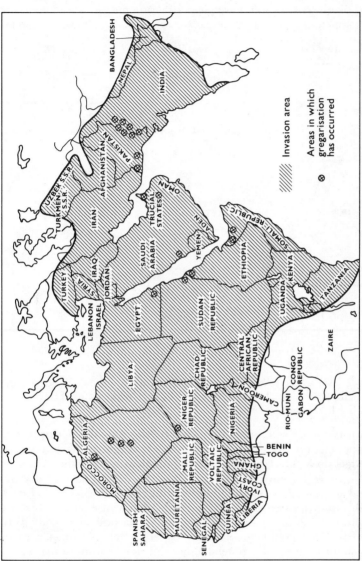

Fig. 6-4 The invasion area of the desert locust and areas in which gregarisation is known or believed to have occurred. (After WALOFF, 1966.)

Ethiopia. Smaller upsurges are probably common elsewhere. However, *Locusta* differs from *Schistocerca* in that only one region, the middle Niger flood plains, is known to have given rise to populations leading to the development of a plague in the last 50 years (Fig. 6–5), and this region is

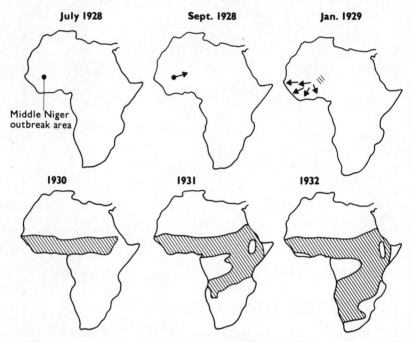

Fig. 6–5 The spread of *Locusta* from its outbreak area on the Middle Niger at the start of the last plague. Arrows show main swarm movements, hatched areas are those heavily infested. (After BATTEN, A., unpublished.)

known as the outbreak area of *Locusta migratoria*. As with *Schistocerca*, very high populations are necessary to produce swarm breeding and only in the Niger do such large populations commonly arise. Here the locusts are able to breed for most of the year because they do so during the rainy season and also as the floods recede from the plains, leaving behind areas suitable for oviposition and subsequent development. In this way *Locusta* can produce four generations in a year so that the potential suitability of the region for population increase is very great. No other area of comparable size has this potential, but this does not mean that under exceptional circumstances a plague could not arise elsewhere.

Nomadacris is similar to *Locusta* in that there are several regions in east, central and western Africa where populations have increased and band formation occurs, but only two of these, the Rukwa Valley in Tanzania and the swamps surrounding Lake Mweru in Zambia are known to have been implicated in the start of the last plague. *Nomadacris* has only a single annual generation and it is not clear how these areas differ from other regions where upsurges occur. There is no evidence that breeding success is greater and it may be that the form and disposition of the vegetation in the outbreak areas has been such as to permit more effective gregarisation.

There does not seem to be any fundamental difference between *Schistocerca* and the other locusts, except that it does not have the permanent, well-defined outbreak areas of the other species. Increases in the desert locust population which lead to gregarisation and swarms formation may occur in many areas in a broad belt from West Africa to India (Fig. 6–4), but because of the very arid habitat in which *S. gregaria* occurs these areas will generally only be suitable for successful breeding at infrequent and irregular intervals.

7 Ecology and Population Fluctuations

7.1 General ecology

The gross ecology of the different locust species varies considerably. *Locusta*, for instance, is widespread in tropical grasslands and its outbreak area is the flood-plain of the middle Niger; *Schistocerca gregaria*, on the other hand, lives in the non-swarming phase in semi-desert regions; while the Sahelian tree locust, *Anacridium melanorhodon* is found in woodland savanna in which species of *Acacia* predominate. But all the species have in common two basic ecological requirements for their development: vegetation for food and shelter, and bare ground for oviposition. Because of this duality of ecological requirements, locusts and many other acridids occur most abundantly in areas with a mosaic of different vegetation types; areas with a uniform plant cover are not generally favourable. With locusts the mosaic of vegetation provides not only a suitable habitat for reproduction and development, but also conditions which favour concentration of the insects, an essential preliminary to gregarisation (p. 45).

Vegetation mosaics commonly develop in zones of contact between two types of plant cover and a characteristic feature of these zones is their instability, one or the other component being favoured by the existing environmental conditions. This instability seems to be closely correlated with locust outbreaks. With *Locusta*, for instance, the drying out of the flood plains results in the development of 'islands' of vegetation in which concentration and successful breeding can occur. With *S. gregaria*, rain produces the flush of vegetation necessary for successful development, but the progressive drying out of the habitat will produce the patchy vegetation in which concentration occurs.

In some cases the vegetation mosaics have developed as a result of human activity. In Australia, for instance, clearing woodlands for pasture has favoured short, tussocky grasses which provide food for the plague locust, and heavy grazing by cattle, sheep and rabbits resulted in an enormous extension of the mosaic habitats suitable for locust development, especially in less arid grasslands which were previously unsuitable for the species. Overgrazing has also produced habitats favouring the development of the South African brown locust. Elsewhere, in Indonesia, deforestation has provided a habitat for *Locusta* where previously none existed. Associated with agricultural methods which involve shifting cultivation and grass burning, conditions have been created which favour the development of swarms, and a similar sequence of events has resulted in considerable increase in the importance of the Bombay locust in Thailand.

7.2 Reproductive potential

Locusts potentially have a very great capacity for increase. A solitary female of the red locust, for instance, has about 170 ovarioles and may lay six egg pods. Assuming a 50:50 sex ratio a single female is potentially capable of producing 1020 offspring in the first generation, 520 200 in the next and 265 302 000 locusts by the third generation. In practice this does not occur. In a non-swarming period the population will fluctuate about an average low level. This demands that each female gives rise, on average, only to one male and one female, so the actual rate of reproduction falls far short of the potential. But there are times when the potential reproduction rate is more nearly achieved and under these conditions an upsurge occurs in the population which may lead to a plague. Why is it that the usual rate of multiplication is so low and how is it that sometimes much higher rates are achieved?

The potential multiplication rate is reduced by a failure to produce the expected number of eggs, by mortality in the eggs and by post-embryonic mortality. All of these may be influenced by other living organisms, the biotic factors, or by the physical characteristics of the environment, though the distinction between the two blurs when factors such as availability of food are considered.

Solitarious locusts are hard to find and hence to study. As a result most studies have been concerned with swarming populations, but because these are mobile and very large they present a very different problem from the more static low density populations and conclusions drawn from swarm populations may not be relevant to non-swarming populations.

7.3 Fecundity and fertility

These terms are often confused, but here fecundity is taken to be the total reproductive potential of a female locust, while fertility is the number of viable eggs produced.

The number of eggs laid in a pod is invariably less than the ovariole number (Table 3, p. 43). It is not at all clear why the insect always falls short of its potential fecundity, but it is known that much of this failure can be attributed to resorption of oocytes, that is the oocytes start to develop in a normal way, but development stops and the yolk already laid down is digested and resorbed into the haemolymph. Poor food supply increases the amount of resorption, and if the food is poor from the beginning of adult life the process of vitellogenesis (yolk deposition) may not begin at all.

The number of egg pods produced by a female locust in the field is difficult to assess, but a good deal of work has been done especially on the red locust. In some seasons mature females may die before laying any eggs at all, but in two seasons out of seven which have been studied, most females laid two pods; only rarely were as many as five egg pods laid and in most years each female only laid, on average, one pod (Table 4).

Table 4 Number of egg-pods laid by female red locusts in parts of the Rukwa Valley outbreak area in different years, expressed as the percentage of females laying the stated number of pods

Year	Number of egg pods						Average number/female
	0	1	2	3	4	5	
1952–3	0	0	80	15	5	0	2.3
1953–4	40	53	7	0	0	0	0.7
1954–5	0	0	50	0	45	5	3.1
1955–6	30	50	18	2	0	0	0.9
1959–60	0	60	37	3	0	0	1.5
1960–1	0	50	42	8	0	0	1.7
1961–2	0	78	22	0	0	0	1.5

It is not clear why the females should die without laying the full complement of eggs, but it is apparent that these two phenomena, oocyte resorption and adult mortality, greatly reduce the potential for increase from generation to generation. In the best season, 1954–5, each female red locust laid, on average, 3 pods, each containing 120 eggs, so that she might have produced 360 offspring (compared with a potential 1020); in the worst season, 1953–4, the average number of pods laid per female was less than 0.7 so that the maximum number of offspring produced per female would have been 84.

Female fertility is clearly a factor of great importance in regulating the multiplication rate of red locusts and this has also been found with the migratory locust in Mali. But even in the worst years the multiplication rate, in terms of viable eggs produced, exceeds 30 so that other factors, operating later in the life cycle, are also important.

7.4 Egg mortality

Not all the eggs laid by a female are fertile, but the proportion of eggs in field populations which fail to develop is generally low, less than 10%. Subsequently, however, during embryonic development eggs are subject to attack by a variety of parasites and predators and are exposed to the vicissitudes of the physical environment.

Parasites and predators are not very important in controlling low density populations. A common parasite is a small wasp of the genus *Scelio*. The female wasp digs into the froth of the egg-pod and lays one egg in each host egg which she encounters and as she has a capacity of over 100 eggs it is common for nearly all the locust eggs in a pod to be parasitized. Usually the number of egg-pods affected is low, 1–3%, but as many as 13% of *Locusta* pods have been found parasitized in the Mali outbreak area. Predacious bettle larvae and rats have also been noted attacking locust eggs in the outbreak areas, but their effects are not, apparently, significant.

In contrast, predatory insects are commonly recorded causing enormous damage in the egg fields of swarms of the desert locust. The most important are the larvae of a fly called *Stomorhina lunata* which belongs to the family Calliphoridae and looks rather like a housefly. This insect seems to be attracted to locust swarms about to oviposit. The female fly pushes her abdomen into the froth at the top of the egg-pod plug as soon as the locust finishes ovipositing and she lays 1–10 eggs. The fly larvae feed on the locust eggs by puncturing the chorion and sucking out the fluid contents. Usually the whole pod is destroyed; eggs which are not eaten are destroyed by the movements of the larvae. In the cases investigated the overall mortality due to this fly amounts to about 20%, but in some cases it reached 100%. *Stomorhina* is regarded as the most consistent cause of mortality in egg fields of *Schistocerca* in east Africa. Egg-pods of the other locusts are also attacked.

Another fly, *Systoechus* and beetles of the genus *Trox* are other important predators of desert locust eggs. They may destroy up to 75% of the eggs, but are less consistent in their occurrence than *Stomorhina* and so are generally less important.

The eggs of *Nomadacris, Schistocerca* and *Locusta* normally develop without any delay in their development. They require water for this development and are not well protected against loss of water. Hence if they are subjected to severe drying conditions, even for a few days, they desiccate. This happened with over 50% of the eggs in the Rukwa red locust outbreak area in 1960–1. Heavy rain fell soon after the eggs were laid, but this was followed by 20 very hot days with hardly any rain at all. Too much water also kills eggs, probably because they are unable to respire, and in 1961–2 waterlogging of the soil due to prolonged heavy rain seems to have been a cause of much mortality to red locust eggs. Clearly for big populations to develop the appropriate distribution of rainfall is essential.

7.5 Post-embryonic mortality

It is generally believed that many locusts of all species die soon after hatching, and studies on the red locust in the Rukwa outbreak area suggest that very high mortality occurs in the first three nymphal instars, but that subsequently, relatively little mortality occurs. Up to 30% mortality of newly hatched nymphs of the desert locust in bands has been seen to result from cannibalism, despite the presence of an ample supply of food plants. First instar nymphs which had started to feed were seen eating newly hatched nymphs as they emerged from the ground. Such cannibalism was observed on every occasion on which hatching was watched carefully, but it was only occasionally seen in later instars. This is a factor of some importance in gregarious populations; it can have no significance in low density populations where the egg-pods are widely separated from each other and the nymphs from one pod tend to hatch within a short period.

Some insects die because they do not moult properly or are attacked by other insects during the process, but mortality as a result of this is not likely to exceed 4%.

Parasites are unimportant in the red locust, but in static bands of the desert locust 20–30% of the nymphs may be parasitized by the larvae of a fly, *Blaesoxipha*. These flies do not lay their eggs in the usual way; the eggs hatch inside the parent fly and larvae are deposited directly on the host locust (the fly is said to be larviparous). One or more fly larvae enter the host through an intersegmental membrane and feed on the haemolymph and fat body of the host. When fully developed they escape from the host, generally through the dorsal surface of the neck membrane, and pupate in the soil. Sometimes the hosts recover, but usually they die, apparently due to desiccation. Although the same species of *Blaesoxipha* parasitizes locusts and grasshoppers it only becomes an important controlling agent when the host populations are dense and static. Various other flies and parasitic wasps are parasites of locust nymphs and occasionally they are present in large numbers, but their occurrence is so spasmodic that they can be of no general significance in controlling the populations.

Locust nymphs are eaten by a wide range of predators and in the Rukwa outbreak area of the red locust it is suggested that robber flies (Asilidae) and dragonflies are very important. Spiders and vertebrate predators also eat locusts, but are unlikely to be significant when the locusts are in low densities and so no more abundant than many other insects. When the locusts are in bands, however, they present easy targets and predators may eat very large numbers. Birds are particularly important because they have the mobility to keep up with the movements of the locusts, but despite their enormous capacity (a stork may eat 500 nymphs in a day) their effect on the population is not normally regarded as very significant. In one instance which is well documented, predators were shown to account for only 8% of the recorded mortality. On the other hand it is recorded that small bands of desert locust nymphs resulting from scattered egg laying were completely eliminated by predators. Larger bands were also destroyed when restricted rainfall caused the locusts and predators to be concentrated in the same area.

Very occasionally gregarious locust populations may be almost totally destroyed by fungus epidemics. The development of these epidemics depends on appropriate conditions of moisture and warmth, but once established they sweep rapidly through the population which is highly susceptible because of its gregarious habit. One account of an epidemic of *Metarrhizium anisopliae* in a desert locust swarm records that the swarm was largely destroyed. Counts of over 50 dead insects per bush were common. The same fungus has also been recorded from the red locust. It has been estimated that in two seasons the fungus *Entomophthora grylli* killed over 90% of adult *Patanga* in parts of Thailand, but in a third, drier, year the fungus was not seen.

Lack of food does not normally appear to be a cause of mortality amongst locusts, although there will certainly be situations where food shortage does occur. Shortage of water, resulting from lack of moist food, may lead to cannibalism of apparently healthy nymphs in mid-instar, but it is not known how general this is, nor what the overall effect on the population would be.

7.6 Population fluctuations

The main factors affecting population size are listed in Table 5. It is apparent that, apart from egg production, biotic factors are not of great

Table 5 Major limitations on population increase in low density and swarming locust populations

+++ indicates a common source of loss of productivity, ++ occasionally important, + rarely important, ● probably not important

Stage	Loss of productivity due to	Low density	Swarming
Adults	egg resorption	+++	+++
	lack of pod production	+++	+++
Eggs	parasites	●	·++
	predators	●	+++
	unsuitable incubation conditions	++	++
Nymphs	cannibalism	●	++
	parasites	●	+
	predators	●	++
	epidemics	●	+

significance in limiting major fluctuations in non-swarming, low density locust populations. This does not mean that they have no effect, but that their effect is normally small. The important factors appear to be the numbers of eggs produced and the survival of these eggs as affected by incubation conditions.

Since egg production by the adults is to a large extent dependent on the availability of suitable food, and successful incubation of the eggs depends on an appropriate soil moisture content, the amount and distribution of rain during the period of reproduction is of key importance.

This is illustrated for the red locust in the Rukwa outbreak area by the equation:

$$y = 6.518 - 0.160X_1 + 0.425X_2 + 0.092X_3$$

where y = the size of adult population.

X_1 = the average rainfall in the catchment area during the *previous* rainy season.

X_2 = the size of the parental (breeding) population.

X_3 = the rainfall in the Rukwa valley during the breeding season (October to December).

The size of the breeding population (X_2) is obviously important because this determines the maximum number of eggs which can be laid. Rainfall during the breeding season (X_3) is important because it provides good vegetation for the breeding adults, moist soil for incubation and vegetation for the newly hatched nymphs. The negative correlation with rainfall in the catchment area during the previous season (X_1) may arise through an effect on the level of the water table. High rainfall in the catchment area will raise the level of the water table and so, if X_3 is also high, lead to waterlogging of the eggs.

Locusts in swarming populations are subject to many different pressures (Table 5). These will not all occur together, nor will they occur consistently, but their effect will be constantly to depress the population. That plagues persist at all is a measure of how successful breeding may be in maintaining the population in spite of these pressures.

The fluctuations in population size which result in plague development and decline are enormous. No attempt has ever been made to express them in terms of numbers of insects because adequate data are not available, but an indication of the changes which have occurred can be obtained from the numbers of countries reporting swarms. Both the red locust and the migratory locust have had one major plague period during this century (Fig. 7–1) beginning in the late 1920s. The desert locust has gone through four plague periods in the same time and perhaps this reflects its widespread distribution with increased chances of conditions leading to a sustained outbreak.

For a general review of locust and grasshopper population dynamics see DEMPSTER, 1963. For an account of the population dynamics of the red locust see STORTENBEKER, 1967. For biotic factors affecting the desert locust see GREATHEAD, 1966. For insect enemies of locusts see GREATHEAD, 1963.

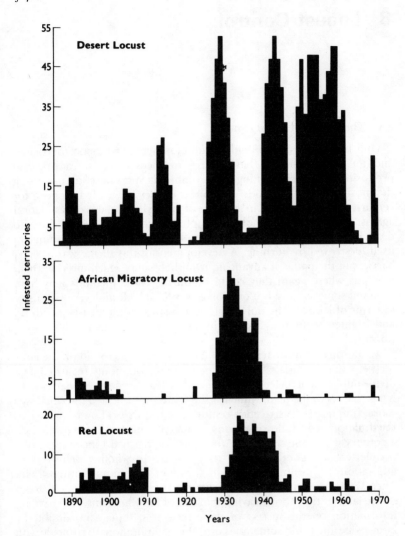

Fig. 7–1 Infestations of the three main African locust species from 1890 to 1970. The level of infestation is indicated by the number of territories reporting swarms. (From THE LOCUST HANDBOOK.)

8　Locust Control

8.1　The origins of locust control

Only in very recent years with the development of synthetic insecticides and the use of aircraft has control of locusts become a practical proposition. At the beginning of the century farmers were still practising the most primitive methods of control, as they probably had been for centuries before. Control, in the sense of killing locusts, was only practical against the relatively immobile nymphal stages. These were beaten to death by gangs of people using sticks and branches or, if the situation lent itself, destroyed by burning. A development of this theme was to erect barriers in the paths of advancing nymphal bands which directed them into pits where again they could be burnt. These methods may have achieved some success, but only on a very local scale and were expensive in terms of labour. The only protection against adult locusts was to try and frighten them away by lighting fires or making as much noise as possible.

An alternative approach, for the less practical, was to call on the gods for help since locusts, and other pests, were commonly regarded as a divine punishment meted out on the sinful. The simplest approach was to hang charms in the field, while for the more sophisticated the ritual was more complicated. A sixteenth-century monk describes how he dealt with the situation in the following terms: 'Thus chanting [psalms] we went into a country where the corn was; which, having reached I made them [the people] catch a good many of these locusts to whom I delivered an adjuration . . . summoning, admonishing, excommunicating them. Then I charged them in three hours time to depart to the sea or else to go to the land of the Moors leaving the land of the Christians.' We have no information on the success of these measures, though no doubt the knowledgeable could sometimes time their admonitions to coincide with the departure of the swarms.

But it is only with the use of poisons that locust control became a practical proposition. The earliest poison used was sodium arsenite, but this was subsequently replaced by benzene hexachloride (BHC) and then by other synthetic insecticides. Arsenical poisons were only effective if eaten by the locusts and so had to be mixed with bait material of some kind which provided the food. The most effective bait was wheat bran and this was used either with or without water. Although effective, baiting was expensive because of the cost of the bait material and its bulk which had to be transported to remote regions for use. For instance, between October and December 1953, 1000 tons of bait were used against the desert locust in east Africa. It required 83 vehicles and 1000 men to find the locust bands and apply the bait, yet of the 1000 tons, the poison (BHC) weighed only 1.3 tons. Clearly the direct application of poison is cheaper and baits are no longer generally used, but it is possible that they may have some role in the future because baits are more specific in their effects than blanket sprays of insecticides.

8.2   Modern methods of chemical control

Insecticides are now widely used in the control of locusts, but some insecticides are many times more effective than others in killing locusts (Table 6) and these belong to different classes of insecticide. The selection

Table 6   The effectiveness of different insecticides against locusts and their mammalian toxicities. The LD_{50} is the dose which when applied to a group of animals kills 50% of them

Class of compound	Name	Toxicity to locusts (LD_{50} in μg/g)		Toxicity to rats (LD_{50} in mg/kg)	
		Schistocerca	Locusta	oral	dermal
Organochlorine	γ-BHC	9	2–7	90	900
	dieldrin	5	2	46	10–102
	DDT	100	100	115	2500
Organophosphate	fenitrothion	5	2	250	200
	malathion	31	24–48	2800	4100
	parathion	2	1	4–13	7–21
Carbamate	carbaryl	25–37	—	850	4000

of an insecticide for practical use depends also on its other properties, especially its persistence, on its mammalian toxicity and on its cost. Parathion, for instance, is very effective in killing locusts, but it is also highly toxic to man (its median lethal dose. LD_{50}, is low) so that it is not regarded as a safe insecticide to use in locust control.

Insecticides may be used to kill locusts in one of two ways, by direct contact with the outer surface and subsequent absorption through the cuticle or by ingestion with the food and absorption via the gut. In the first case the insecticide needs to be highly toxic to locusts, but persistence is of no advantage. In the second, persistence is essential because it is sprayed onto vegetation likely to be eaten by the locusts; the longer it persists the more locusts it is likely to affect. This method can only be used against locust populations that are relatively static; with flying swarms a contact, knock down, poison is necessary. The insecticides most commonly used against locusts are BHC and dieldrin, although there is a continuing search for other, safer insecticides.

Sometimes insecticides are applied as a dust, but it is much more usual to apply them in solution or as an emulsion in water or oil using a mechanical pump or sprayer. There are two different techniques of spraying: target spraying in which the spray is aimed directly at the locusts, or drift spraying in which use is made of the wind to drift the insecticide over the general area occupied by the locusts. The first method has the advantage that it can be employed at any time and that the dispersal of poison can be limited, but it has the great disadvantage that a thorough search for the locusts is necessary in order to locate all the bands and in difficult terrain some bands prove to be totally inaccessible. In control operations against red locust nymphs an average of only 4.7 ha/day was controlled by this direct application. With a machine using the drift principle, on the other hand, an average spraying rate of 12.5 ha/h was achieved, but the blanket spraying of large areas is only justified if the populations are dense and fairly continuous. With drift spraying from the ground, as with control from aircraft, the size of droplet produced by the sprayer is critically important. Too large a drop of poison solution falls to the ground too quickly to drift far; too small a drop, on the other hand, is likely to be carried up on rising air currents and lost entirely. So the size of the drop has to be regulated by using an appropriate spray nozzle.

A potentially more efficient means of controlling locust nymphs, but one that has been little tried, is by barrier or lattice spraying. The principle is to spray a series of strips of vegetation with a persistent poison which is eaten by locusts as they move through them. The strips may be transverse to the line of movement of the locusts or in the form of a lattice and the spacing of the strips depends on the rate of movement of the locusts. A compromise must be reached between having the strips a long way apart, so reducing the cost of the operation, and having them close together so that even relatively immobile groups of insects move into the sprayed strips. It is also important that that the amount of poison put onto the vegetation and the width of the sprayed strips is such as to ensure that the locusts feed while in the strips and that they ingest lethal amounts of poison with their meal. In a trial against the red locust successive spray lines were 500 m apart and the spray swaths about 20 m wide, while successful control of desert locust nymphs has been achieved with sprayed strips 2 km apart. The lattice may be applied from the ground or the air. No very extensive trials have been carried out using this method, but effective control of the desert and red locusts has been achieved using dieldrin, which persists for 40 days or more.

Nymphal bands can be controlled from the ground relatively efficiently; adults are much more difficult because they can fly away. Against adult red locusts drift spraying from the ground was partially successful, but work could only be carried out at night when the locusts were in the roosts. This difficulty, together with the great cost of maintaining ground spraying machinery in inaccessible areas, has led to an increasing use of aircraft in locust control. Against bands of nymphs, aircraft can put down an insecticide barrier much more quickly and effectively than ground machinery. Aircraft can also attack settled adult swarms, but it is in attacking flying swarms that their real value is seen. Flying locusts present a bigger target than settled ones so that they are more likely to be hit by drops of poison, although the density of locusts in flight is very much less than their density on the ground. For instance a swarm which in flight varied in area from 130 to 4650 ha occupied only 4 to 520 ha when on the ground (the wide variation arose because estimates were made on successive days in very different habitats). This takes no account of the volume density which depends on the height of the ceiling at which the insects are flying. Density will obviously be greater when swarms are low flying and the best time for spraying is soon after the locusts have taken off.

In aerial spraying it is an advantage to use relatively small drops of poison because these remain in the air for longer and so have a better chance of hitting a locust. Another reason for attacking flying rather than settled locusts is that in a settled swarm many of the small droplets impact on the vegetation and so are effectively lost; in the absence of vegetation these droplets remain in the air for longer and so have an increased chance of hitting a locust. But a disadvantage of the small droplets is that, unless a highly toxic insecticide is used, locusts may receive sublethal doses of insecticide. For this reason it is necessary to select insecticides which are contact poisons and which are cumulative in their effects, otherwise the receipt of sublethal doses of poison will tend to select for insects which are more resistant to the insecticide. That locusts can develop insecticide resistance has been shown in laboratory trials, and over four generations the LD_{50} may be more than doubled. In other words more than twice as much poison would be needed to achieve an equal kill or it would be necessary to switch to a different insecticide, but even here there are problems because the development of resistance of the desert locust to BHC also confers some degree of resistance to dieldrin and fenitrothion.

Resistance of locusts to insecticide is not known to have occurred in the field and selection is unlikely to be intense because of the mobility of locusts. Only a small percentage of any generation is attacked so that the chances are low of two successive generations receiving sublethal doses and building up resistance. It is, however, possible that this could happen in the outbreak areas where control is repeatedly carried out.

8.3 The strategy of locust control

Because locusts are so highly mobile and because they can destroy a crop very rapidly, normal methods of crop protection are inadequate. There is not time to send for a control team after the locusts have arrived because by the time the team is ready for action the crop will have been destroyed. The locusts must be attacked before they reach the crop. Two approaches to the problem are possible which largely reflect the ecological differences between the desert locust on the one hand and the migratory and red locusts on the other.

The migratory and red locusts have well-defined outbreak areas. If locust swarms are prevented from leaving these areas the chances of a widespread plague developing are reduced to a very low level. This kind of preventive control entails the continual surveillance of the outbreak areas to detect local upsurges, and the control of the locusts whenever high numbers occur. Ideally swarms are prevented from forming at all, but if they do, every effort is made to contain them in the outbreak areas. Since these are to a large extent uninhabited the damage done by the locusts within them is of little importance. With the desert locust the problem is much more difficult. Spreading over an enormous area of North Africa, the Middle East and the Indian subcontinent, and with no well-defined outbreak areas, preventive control is less easy, but regional locust control organizations have teams of locust officers whose task it is to survey the potential breeding areas of the locust at appropriate intervals. In this way incipient outbreaks are discovered and can be controlled.

Thus a major element in the effective control of any of the locust species, but especially the desert locust, is that of obtaining information on the whereabouts of locusts and transmitting this information to the relevant authorities in an appropriate form. Much effort goes into locust survey and information services, for without them control can never be efficient.

8.4   The organization of locust control

The major locust species range over many countries and efficient control demands co-ordinated action within these countries. The initiative for the development of an international system came in the 1920s from B. P. Uvarov, then a Senior Assistant in the Imperial Bureau of Entomology in London, and subsequently the first Director of the Anti-Locust Research Centre. London was an obvious choice as a communications centre because many of the countries involved were at that time British Colonies and lines of communication with London were often better than cross country links. This was clearly recognized at the First International Locust Conference in Rome in 1931 when Uvarov's 'unit' (there were only two people in it!) became generally recognized as the institution responsible for co-ordinating information on all species of locusts in Africa and south-western Asia. This function continued and expanded when, in 1945, the unit became the Anti-Locust Research Centre. Members of the unit's staff were engaged in research and advisory work which led, among other things, to the establishment of the Central African Red Locust Control Organization (later the International Red Locust Control Service) in 1941 and the Organisation Internationale Contre le Criquet Migrateur Africain in 1948, and the production from the early 1940s of a monthly summary of the desert locust situation with a forecast of probable future developments. Such an early warning system is vital for the proper mobilization of the survey and control teams. Subsequently, in 1958, a Desert Locust Information Service was established, funded partly by FAO (the Food and Agriculture Organization of the United Nations) who initiated a Desert Locust Programme in 1952. The international co-ordinating role of the Anti-Locust Research Centre was progressively taken over by the United Nations body, who are now fully responsible, but the Anti-Locust Research Centre (now the Centre for Overseas Pest Research) maintains an advisory and research role.

The International Red Locust Control Service (IRLCS) and the Organisation Internationale Contre le Criquet Migrateur Africain (OICMA) were established to carry out preventive control against the red and migratory locusts respectively. Both were internationally financed organizations, member countries being those in which outbreak areas occurred or which were liable to invasion during plague periods. The IRLCS is based in Zambia and OICMA in Mali. Initially both organizations maintained research and control teams in the outbreak areas, but the cost of this has become increasingly prohibitive since these outbreak areas are remote and the maintenance of permanent teams entails making and maintaining roads, and maintaining camps and supplies. The high cost, and the increasing pressure to reduce costs, has led more and more to reliance on light aircraft, which, while they may be good from the control aspect, are much less effective for the surveying for locusts which is an essential part of the organizations' work. Political problems add to the difficulties. Many of the countries which originally contributed have changed their status and name, some are unable to contribute because of their own financial problems, others are not politically acceptable as members of the organizations by a majority of contributors. And so in 1975 these organizations are in severe financial difficulties and the international character of the locust problem is emphasised once again.

The desert locust does not lend itself to control by a single organization; from the outset individual regions and countries have maintained their own control services. In Africa the two principal regional organizations are the Desert Locust Control Organisation for Eastern Africa based in Ethiopia, and the Organisation Commune de Lutte Antiacridienne et de Lutte Antiaviaire (in West Africa). FAO has five regional co-ordinating centres, in Algiers (for N.W. Africa), Dakar (for W. Africa), Addis Ababa (for E. Africa), Jeddah (for the Near East) and Tehran (for S.W. Asia), and co-ordinates activities between the regions through its headquarters in Rome.

In addition to its general co-ordinating role, FAO has established a Desert Locust Emergency Fund to provide speedy assistance to governments and regional organizations in the early stages of locust outbreaks, and a Trust Fund, to which many nations contribute. The Trust Fund is devoted to a long term programme of research into desert locust survey and control methods. Training schemes are organized with the aim of improving the effectiveness of survey and control, using the most generally accepted and up to date methods, in all parts of the desert locust invasion area.

Other locust species, such as the brown locust, the Bombay locust, and the Australian plague locust are controlled by national organizations since they are almost entirely confined to single countries. The problems of survey and control are similar to those that have already been discussed, but political difficulties do not usually occur.

For a history of the Anti-Locust Research Centre see ROFFEY, 1970. For the development of red locust control see GUNN, 1960. For the role of FAO in locust control see SINGH, 1972.

8.5   Biological control

Locusts have many natural enemies (see p. 52), but, while these may contribute to the natural regulation of locust numbers, their practical use in locust control is unlikely for three reasons: the locusts come in vast numbers, they are highly mobile and their occurrences are irregular. For parasites or predators to be effective they must either be released into the locust population in vast numbers or their populations must build up rapidly in order to combat the enormous locust population. The first proposition is not practical; there is at present no possibility of producing parasites or predators in sufficient numbers or of keeping them alive until locusts appear. The mobility of the locusts militates against the build-up of parasites within the population and there are no proven instances of parasites migrating with swarms, though this has been said to occur with the predatory wasp, *Sphex*.

The use of micro-organisms in biological control of locusts seems a more practical proposition. Micro-organisms such as the fungus *Metarrhizium* can be mass-produced and stored and then applied in the same way as insecticides. They have the potential advantage over insecticides that they are more specific and they are self-regenerating so that they spread progressively through the population. But all attempts to use fungi as controlling agents have failed because the fungal spores will only germinate under specific conditions of moisture and temperature. Since these conditions cannot be controlled, the practical use of fungi does not seem possible. Protozoa are known which kill locusts and their development is less dependent on environmental conditions. On the other hand there is only one verified instance of a natural protozoan epizootic markedly affecting a locust population. This was an outbreak of *Malameba locustae* which prevented an expected outbreak of the brown locust in South Africa.

The most hopeful approach is long-term ecological control. This can only apply to these species with well-defined outbreak areas, but in them an appropriate modification of the habitat could certainly prevent the build-up of locust populations. Some efforts at modifying the Lake Rukwa habitat were made in 1949–59, but they were on too small a scale and were not successful. Large scale modifications are limited by cost. There is no doubt that appropriate control of the level of water in Lake Mweru could eliminate the surrounding plains as important breeding grounds for the red locust, but even in 1959 the cost was prohibitive. The control of flooding from the Niger in Mali would almost certainly destroy the existing flood plains as an outbreak area for *Locusta* but the cost would be very great indeed and many countries would be affected.

Two approaches are possible in future locust control: one involves the refinement of existing techniques and the application of modern technology such as the use of satellites to locate areas where rain has fallen and in which breeding may occur; the other requires a new approach to the problem with the introduction of large scale changes in the locust habitats associated with agricultural development.

The first entails continued high levels of expenditure and problems in international co-operation, the second demands massive initial financial outlay, and again international co-operation, but with more positive returns for the expenditure. It is a sad reflection on our society that we shall probably have to wait for another series of massive locust plagues before politicians and financiers will take a serious, long-term look at the problem.

Further Reading

The key reference work on locusts is *Grasshoppers and Locusts* by B. P. UVAROV (Cambridge University Press). The first volume, published in 1966, deals with anatomy, physiology, development and phase polymorphism. The second volume, scheduled for publication in 1976, covers the behaviour and ecology of locusts and gives accounts of the biology of the major species. A concise, practical account of locusts is given in *The Locust Handbook* published by the Centre for Overseas Pest Research. Only the specialist student will need more information than these books contain.

The following list contains up to date references on the topics as indicated in the text:

ALBRECHT, F. O. (1956). *The Anatomy of the Migratory Locust*. Athlone Press, London.

BERNAYS, E. A. (1971). *Z. Morph. Tiere.*, **70**, 183–200.

BLANEY, W. M. (1974). *J. exp. Biol.*, **60**, 275–94.

BURROWS, M. (1975). *J. exp. Biol.*, **62**, 189–219. *discrimination of chemic by locusts*

CAMHI, J. M. (1970). *J. exp. Biol.*, **52**, 519–31.

CARLISLE, D. B., ELLIS, P. E. and BETTS, E. (1965). *J. Insect Physiol.*, **11**, 1541–58.

CHAPMAN, R. F. (1974). *Feeding in Leaf-Eating Insects*. (Oxford Biology Readers) Oxford University Press, London.

DEMPSTER, J. (1963). *Biol. Rev.*, **38**, 490–529.

DETHIER, V. G. (1963). *The Physiology of Insect Senses*. Methuen, London.

ELLIS, P. E. (1972). *Proc. Int. Study Conference on Current and Future Problems of Acridology*, 63–77.

ELLIS, P. E. and ASHALL, C. (1957). *Anti-Locust Bull.* no. 25; 94 pp.

GREATHEAD, D. J. (1963). *Trans. R. ent. Soc. Lond.*, **114**, 437–517.

GREATHEAD, D. J. (1966). *J. appl. Ecol.*, **3**, 329–50.

GUNN, D. L. (1960). *J. ent. Soc. sthn. Africa*, **23**, 65–125.

HASKELL, P. T. (1966). *Symp. R. ent. Soc. Lond.*, **3**, 29–45.

JOHNSON, C. G. (1969). *Migration and Dispersal of Insects by Flight*. Methuen, London.

LEWIS, C. T. (1970). *Symp. R. ent. Soc. Lond.*, **5**, 59–76.

NORRIS, M. J. (1968). *Colloq. int. Cent. nat. Rech. scient.* no. 173, 147–61.

PENER, M. P., GIRARDIE, A. and JOLY, P. (1972). *Gen. comp. Endocr.*, **19**, 494–508.

PRINGLE, J. W. S. (1965) in *The Physiology of Insecta*, ed. M. Rockstein. Academic Press, New York and London.

RAINEY, R. C. (1963). *Anti-Locust Mem.* no. 7; 115 pp.

ROFFEY, J. (1963). *Anti-Locust Bull.* no. 39; 32 pp.

ROFFEY, J. (1970). *The Anti-Locust Research Centre. A Concise History to 1970.* Anti-Locust Research Centre, London.

ROFFEY, J. and POPOV, G. (1968). *Nature, Lond.*, **219**, 446–50.

SCHAEFER, G. (1976) *Symp. R. ent. Soc. Lond.*, **7**, 157–197.

SINGH, G. (1972). *Proc. Int. Study Conference on Current and Future Problems of Acridology*, 475–85.

SOUTHWOOD, T. R. E. (1962). *Biol. Rev.*, **37**, 171–214.

STORTENBEKER, C. W. (1967). *Agric. Res. Report, Institute for Biological Field Research, Arnhem* no. 894; 118 pp.

WALOFF, Z. (1966). *Anti-Locust Mem.* no. 8; 111 pp.

WALOFF, Z. (1972). *Bull. ent. Res.*, **62**, 1–72.

WEIS-FOGH, T. (1964). *J. exp. Biol.*, **41**, 229–56.